Perspektiven der Mathematikdidaktik

Reihe herausgegeben von

Gabriele Kaiser, Sektion 5, Universität Hamburg, Hamburg, Deutschland

In der Reihe werden Arbeiten zu aktuellen didaktischen Ansätzen zum Lehren und Lernen von Mathematik publiziert, die diese Felder empirisch untersuchen, qualitativ oder quantitativ orientiert. Die Publikationen sollen daher auch Antworten zu drängenden Fragen der Mathematikdidaktik und zu offenen Problemfeldern wie der Wirksamkeit der Lehrerausbildung oder der Implementierung von Innovationen im Mathematikunterricht anbieten. Damit leistet die Reihe einen Beitrag zur empirischen Fundierung der Mathematikdidaktik und zu sich daraus ergebenden Forschungsperspektiven.

Reihe herausgegeben von
Prof. Dr. Gabriele Kaiser
Universität Hamburg

Weitere Bände in der Reihe https://link.springer.com/bookseries/12189

Xiaoli Lu

Novice Mathematics Teachers' Professional Learning

A Multi-Case Study in Shanghai

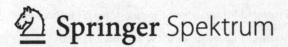

Xiaoli Lu
School of Mathematical Sciences
and Shanghai Key Laboratory of PMMP
East China Normal University
Shanghai, China

Dissertation University of Hong Kong, 2017

ISSN 2522-0799   ISSN 2522-0802   (electronic)
Perspektiven der Mathematikdidaktik
ISBN 978-3-658-37235-4   ISBN 978-3-658-37236-1   (eBook)
https://doi.org/10.1007/978-3-658-37236-1

Responsible Editor: Marija Kojic
This Springer Spektrum imprint is published by the registered company Springer Fachmedien
Wiesbaden GmbH part of Springer Nature.
The registered company address is: Abraham-Lincoln-Str. 46, 65189 Wiesbaden, Germany

Foreword

Teacher quality is the most important element in children's education, and effective teacher professional development (TPD) has been recognised as essential for promoting teacher quality and thus improving students' academic performance. While the literature on mathematics teacher professional development is extensive, focusing on mathematics teacher beliefs, teacher knowledge, teaching practice and environmental influences on TPD, most of this body of literature focussed on only a few components of TPD and the influencing factors, and failed to address the complex and dynamic characteristics of TPD embedded in the mathematics teachers' professional lives and working conditions and the explanatory causality and reciprocal influences of various TPD components.

The present book by Xiaoli Lu, based on her PhD research, aims to fill this research gap. It presents a multiple-case study of novice teachers with the aim of providing key information and insights for TPD. The book's target readership includes future and practicing teachers at all levels as well as researchers, teacher educators and policy-makers. The book presents in detail the professional learning processes of three novice mathematics teachers during the initial two years of their teaching careers in Shanghai. It documents the novice teachers' professional learning outcomes in terms of teacher beliefs, knowledge and teaching practice as well as the opportunities and challenges in their learning of teaching during the TPD process. The findings illustrate the interactions among teacher beliefs, knowledge and teaching practice as well as the influences of environmental factors on the ways in which teachers achieved professional learning. In particular, the three teacher cases provide important implications for teacher educators and policy-makers by promoting early career teachers' professional learning of student-centred pedagogies in contexts dominated by traditional teaching practice.

The multiple-case study followed a rigorous research methodology and design and was based on a comprehensive survey of literature on TPD theories, teacher expertise and mathematics teachers' teaching in the context of China. The book presents in detail the professional learning processes of three novice mathematics teachers who had different learning experiences and taught in different schools. The findings from the three cases were synthesised to examine the opportunities and constraints that various mentorships present in terms of teachers' professional learning. The study's findings reveal how teachers' experiences, beliefs, knowledge and practices interact to produce various learning outcomes and the important role that the environment plays in novice mathematics teachers' implementation of student- or teacher-centred teaching within the context in which they are situated.

Two journal articles have been published based on the study presented in this book. That by Lu, Kaiser and Leung (2020) focuses on a model that examines various approaches to mentoring, while that by Lu, Leung and Li (2021) investigates novice mathematics teachers' agency with respect to integrating history into mathematics teaching in a performance-driven context.

This book and the two above-mentioned articles address a research gap in the field of TPD. Taking the specific subject and context into consideration, this study reveals the complexity, dynamics and openness of TPD in present-day mathematics education. We hope that this book will contribute to enriching theoretical knowledge regarding TPD and to enhancing mathematics teaching in China and beyond.

Frederick K. S. Leung
The University of Hong Kong
Hong Kong, China

Acknowledgements

I wish to thank the numerous people who supported me throughout my completion of the study.

First, I would like to express my sincerest gratitude to my supervisor, Professor Frederick K. S. Leung. Every class, meeting and conversation with him enhanced my learning and promoted my growth as a researcher. His patient and insightful feedback played an essential role in the study's completion. He also provided us with many opportunities to communicate with other scholars and to offer necessary support for my living in Hong Kong. Without his guidance and encouragement, my learning process over the four years would not have been so smooth.

I am also grateful to the three participants and their mentors who participated in the study and without whom there would be no data to discuss. I am also grateful for the assistance offered by Professor Bao Jiansheng, Professor Gu Lingyuan, Professor Wang Xiaoqin, Dr. Gu Feishi, Dr. Zhu Yan, Dr. Cheng Jing, Mr. Zou Jiachen, and Mr. Cui Weixin who helped me to establish contact with the participants.

I would also like to extend my appreciation to the several scholars who provided constructive comments on my study: Dr. Ida Mok, Professor Bao Jiansheng, Professor Gu Lingyuan, Dr. Zhu Yan, Dr. Zhang Qiaoping, Professor David Clarke, Professor Fan Lianghuo and Professor Susan Leung. I also wish to thank my friends who have accompanied me on my academic journey, particularly Ailin, Cathy, Christy, Danial, Jiming, Judy, Lina, Peigen, Tracey Tao, Tracey Xu, Wei, Zhan and Yin. Particular thanks are due to Leming Liang for sharing ideas and discussing with me in detail the issues I encountered when compiling the thesis and to Jiali Tang for re-preparing the figures in the book and helping to check the references prior to publication.

Sincere gratitude is owed to Professor Gabriele Kaiser for her enthusiastic encouragement and deep interest in the study, as well as effective research guidance and collaboration for further exploration of the study after my graduation. I am also grateful to Professor Jianpan Wang for his support in the study's further exploration during my time at the East China Normal University, Shanghai.

In addition, I wish to thank my friends in Hong Kong and Shanghai. Special thanks are due to Rachel and Shengli, who extended their help when I first arrived in Hong Kong, and to Jiao, Di, Lei, Le, Miao, Liang, Haiyan, Gang, Jing, Hongdao, Hongyan, Weiqin, Jialu and Likun who were of assistance when I collected data in Shanghai. I would also like to thank my friends from the badminton team, hiking team, and the Faculty of Education at the University of Hong Kong and the Department of Mathematics East China Normal University, who made the four years colourful and wonderful.

Finally, my sincerest appreciation and deepest love are extended to my parents and my husband, whose support, concern and company are consistently the greatest source sustenance for me.

The book is supported by a grant from the Science and Technology Commission of Shanghai Municipality (No. 18dz2271000)

February 9, 2022   Xiaoli Lu
                                                                                         鲁小莉

Abstract

This study's central aim is to explore how three novice upper secondary school mathematics teachers in Shanghai experienced the professional learning process in the early stage of their careers in situated contexts. Teacher professional learning is considered a complex and dynamic system that connects both cognitive and situated perspectives on learning theory.

The study adopted a longitudinal case study approach in which teachers' beliefs, knowledge and teaching practices were analysed over the two academic years from 2013 to 2015. The data consisted primarily of classroom observation and interviews and were collected in four rounds. In each round, three or four consecutive lessons for each teacher were observed, and semi-structured interviews were conducted that focused on the teachers' background information, beliefs, knowledge and teaching practice, as well as their own reflections on the learning process and their mentors' perceptions of their interactions with their mentees. A qualitative data analysis approach was adopted to generate a holistic description of the teachers' teaching and their pedagogical learning.

Doris, who underwent teaching-related training in her bachelor's and master's programmes and during a one-year voluntary teaching practice, focused on learning to teach school mathematics in a teacher-centred way that was consistent with the collective ideas of other teachers in the same environment while integrating the history and culture of mathematics into her teaching in a bid to promote students' interest and mathematical thinking. Jerry, who learned to teach mathematics during his bachelor's programme and obtained a master's degree in mathematics, in particular delivered performance-oriented learning, as required by the school environment. Tommy, who had not received any teaching-related training, focused particularly on learning how to teach mathematics in a school context and adjusted his original pedagogical beliefs accordingly.

    Combining the three cases revealed that the three teachers brought differ-
ent beliefs and knowledge to their teaching practices owing to their different
individual experiences. However, they consistently implemented teacher-centred,
content-focused and performance-oriented teaching practice over the two years in
the school context and learned related knowledge during the professional learning
process. At the end of the two years, interviews with the three teachers revealed
contradictions between their stated beliefs and their actual teaching. While in
principle they tended to support student-centred teaching, in practice, their teaching focused on the contents and students' performances, as demanded by the
environment. Moreover, the novice teachers experienced seven one-to-one mentorships during the two-year teaching practice period, four of which were found
to be unnecessary, one demonstrative, one mainly demonstrative and sometimes
collaborative, and one mainly supportive and sometimes collaborative. The opportunities and challenges associated with the various mentorships in terms of the
novice teachers' professional development are discussed. The study finally summarises the three teachers' learning outcomes during the two-year professional
learning period as well as the influences on their professional learning.

    The results indicate that the novice mathematics teachers' professional learn-
ing processes differed owing to their different previous experiences in learning
and teaching mathematics and the environmental influences and that the novice
teachers could promote student-centred pedagogies (e.g., integrating history into
teaching); however, novice teachers' professional learning is heavily influenced
by the environmental norms, which reveals a dominant performance-driven context. The study's findings contribute to verifying and enriching the theory of
teacher professional learning, a comprehensive understanding of novice teachers'
professional learning process in a particular subject—here, mathematics—and
within the specific context of China as well as insights into pre-service and
in-service teacher education. Further research is needed to widen the scope of
the investigation with the aim of better understanding and promoting novice
mathematics teachers' professional learning in different environmental contexts.

Contents

1 **Introduction** 1
  1.1 Rationale for the Study 1
  1.2 Aims of the Study 3
  1.3 Overview of Research Methodology 4
  1.4 Significance of the Study 5
  1.5 Overview of the Book 5

2 **Literature Review and Theoretical Framework** 7
  2.1 Teacher Professional Learning 7
    2.1.1 A Delimitation of Teacher Professional Learning 8
    2.1.2 Theoretical Background of TPL: The Two
        Perspectives on Learning Theory 8
    2.1.3 Adopting a Theory that Considers the Complexity
        and Dynamics of TPL 9
  2.2 Novice Teachers 10
    2.2.1 The Stage of Expertise 10
    2.2.2 Features of Novice Teachers in China 11
    2.2.3 Mentoring and Induction 12
  2.3 Individual-Level Orientation to Learning 14
    2.3.1 Teacher Beliefs and Knowledge 14
    2.3.2 Teaching Practice 27
    2.3.3 Interactions among the Elements at the Individual
        Level 29
  2.4 Influences of the School on Teacher Learning 30
  2.5 The Theoretical Framework of this Study 32

3   Research Methodology and Study Design 35
 · 3.1   Research Purpose and Research Questions 35
    3.2   Adopting a Qualitative Research Design 36
    3.3   Case Study Research Model 37
          3.3.1   Choosing a Case Study Approach 37
          3.3.2   Participant Recruitment 38
          3.3.3   Research Context 40
    3.4   Data Collection 43
          3.4.1   Data Collection Procedure 43
          3.4.2   Classroom Observation 45
          3.4.3   Interviews 45
          3.4.4   Document Collection 46
          3.4.5   Field Notes 47
    3.5   Data Analysis 47
          3.5.1   Analysis of the Three aspects in Each Stage 49
          3.5.2   Further Analysis 55
          3.5.3   Analysis Focusing on the Influences of Mentoring 56
          3.5.4   Analysis of Supplementary Data 57
    3.6   Validity and Reliability 57
    3.7   Ethical Considerations 59

4   The Case of Doris 61
    4.1   Introduction 61
    4.2   Background 62
          4.2.1   Doris' Background 62
          4.2.2   School Information 63
    4.3   The Initial Stage of the Two-year Professional Learning 64
          4.3.1   Main Features of Doris's Teaching Practice 64
          4.3.2   Teacher Beliefs 66
          4.3.3   Doris's Teacher Knowledge 67
          4.3.4   Summary and Discussion 69
    4.4   The Medium Stage of the Two-year Professional Learning 72
          4.4.1   Main Features of Doris's Teaching Practice 72
          4.4.2   Beliefs Reflected in Teaching (Enacted Beliefs) 76
          4.4.3   Knowledge in Teaching 77
          4.4.4   Summary and Discussion 79
    4.5   The End Stage of the Two-year Professional Learning 80
          4.5.1   Main Features of Doris's Teaching Practice 80
          4.5.2   Doris' Teacher Beliefs 82

	4.5.3 Doris's Teacher Knowledge	84
	4.5.4 Doris's Self-reflection on Her Learning and Words from Zhao	85
	4.5.5 Summary and Discussion	87
4.6	Doris's Professional Learning	87
	4.6.1 Overall Learning Outcomes	87
	4.6.2 The Ways to Achieve Her Learning Outcomes	88
5	**The Case of Jerry**	**93**
5.1	Introduction	93
5.2	Background	94
	5.2.1 Jerry's Background	94
	5.2.2 School Information	95
5.3	The Initial Stage of Two-year Professional Learning	96
	5.3.1 Main Features of Jerry's Teaching Practice	96
	5.3.2 Jerry's Teacher Beliefs	99
	5.3.3 Jerry's Teacher Knowledge	100
	5.3.4 Summary and Discussion	101
5.4	The Medium Stages of the Two-year Professional Learning	104
	5.4.1 Main Features of Teaching Practice	104
	5.4.2 Beliefs Reflected in Teaching (Enacted Beliefs)	107
	5.4.3 Knowledge in Teaching	108
	5.4.4 Summary and Discussion	109
5.5	The End Stage of the Two-year Professional Learning	110
	5.5.1 Main Features of Teaching Practice	110
	5.5.2 Jerry's Teacher Beliefs	112
	5.5.3 Jerry's Teacher Knowledge	113
	5.5.4 Jerry's Self-reflection on His Learning and Comments on the Mentoring	114
	5.5.5 Summary and Discussion	116
6	**The Case of Tommy**	**119**
6.1	Introduction	119
6.2	Background	119
	6.2.1 Tommy's Background	119
	6.2.2 School Information	120
6.3	The Initial Stage of the Two-year Professional Learning	122
	6.3.1 Main Features of Teaching Practice	122
	6.3.2 Tommy's Teacher Beliefs	123
	6.3.3 Tommy's Teacher Knowledge	124
	6.3.4 Summary and Discussion	126

6.4   The Medium Stages of the Two-year Professional Learning   128
        6.4.1   Main Features of Teaching Practice   128
        6.4.2   Beliefs Reflected in Teaching (Enacted Beliefs)   131
        6.4.3   Knowledge in Teaching   132
        6.4.4   Summary and Discussion   133
6.5   The End Stage of the Two-year Professional Learning   134
        6.5.1   Main Features of Tommy's Teaching Practice   134
        6.5.2   Tommy's Teacher Beliefs   136
        6.5.3   Teacher Knowledge   137
        6.5.4   Tommy's Self-reflection on His Learning and Words
                from Li   138
        6.5.5   Summary and Discussion   140

7   The Mentorship For Doris, Jerry, And Tommy   145
7.1   Introduction   145
7.2   Seven Mentors for Doris, Jerry and Tommy   145
7.3   The Types of Mentorship   147
        7.3.1   Mentor's and Mentee's Input   147
        7.3.2   Mentee's Self-reflection and Attitudes Towards
                Mentorship   150
7.4   The Opportunities and Constraints of Mentorship   152
        7.4.1   General Mentorship Types: A Summary   152
        7.4.2   The Nature of the Four Types of Mentorship:
                A Discussion   153
        7.4.3   Conclusions   154

8   Discussion And Conclusion Of The Study   157
8.1   Discussion of Findings in the Study   157
        8.1.1   Learning Outcomes   158
        8.1.2   The Process of the Two-year Professional Learning   165
8.2   Significance of the Study   178
        8.2.1   Theoretical Significances   178
        8.2.2   Practical Significances   179
8.3   Limitations of the Study   180
8.4   Suggestions for Further Research   182
8.5   Envoi   182

References   185

List of Figures

Figure 2.1   Conceptual framework of the study 33
Figure 4.1   The unlettered graph for the proof of formulae 74
Figure 4.2   Doris' professional learning focused on the teaching
             of school mathematics 90
Figure 4.3   Doris' professional learning focused on the integration
             of the history and culture of mathematics into teaching 92
Figure 5.1   Jerry's professional learning focused
             on the examination-oriented teaching 118
Figure 6.1   The process of decline in the non-traditional beliefs
             Tommy held at the beginning 143
Figure 8.1   A generic model showing the process of professional
             learning 166

List of Tables

Table 2.1   Ernest's (1989) view of mathematics teacher beliefs 17
Table 2.2   Summary of the three levels of mathematics teacher
            belief 20
Table 2.3   The category of knowledge from MKT, summarised
            from Ball et al. (2008) 24
Table 2.4   The Knowledge Quartet framework 25
Table 3.1   Basic information of teacher participants 40
Table 3.2   Data collection schedule 44
Table 3.3   Summary of data analysis 48
Table 3.4   Coding scheme for teacher questioning in class 52
Table 3.5   Teachers' stated beliefs 53
Table 7.1   One-to-one mentorship for the three novice teachers 146
Table 7.2   The types of mentorship for the three novice teachers 153

Introduction

In recent years, the existing simplistic conceptualisation of teacher professional learning has been recognised as a likely cause of teacher professional learning activities' inefficacy owing to its failure to address the ways in which learning is embedded in teaching processes and the complex contexts in which those processes are enacted (Borko, 2004; Clarke & Hollingsworth, 2002; Opfer & Pedder, 2011). Taking complexity and teaching dynamics into consideration, this study investigates novice mathematics teachers in upper secondary schools in Shanghai from cognitive and situated perspectives on learning to assess the professional learning the teachers experienced in the early stages (i.e., the initial two years) of their teaching and how they engaged in learning in their specific contexts. This introductory chapter begins by introducing the rationale for the study. It then clarifies the study's aims, followed by an overview of the research methodology employed. The study's significance and the organisation of the thesis are presented at the end of the chapter.

1.1 Rationale for the Study

As Opfer and Pedder (2011) pointed out, teachers' professional development has been recognised as an important means of improving both teacher and school quality and of attaining quality student learning. However, professional development efforts to date have yielded disappointing results owing to their simplistic conceptualisations of teacher learning, which fail to consider the ways in which learning is embedded in teachers' professional lives and working conditions. As such, Opfer and Pedder (2011) suggested that teacher professional learning is a complex system consisting of three subsystems (the teacher, the school and the learning activity), and that the investigation thereof should focus on explanatory

© The Author(s), under exclusive license to Springer Fachmedien Wiesbaden GmbH, part of Springer Nature 2022
X. Lu, *Novice Mathematics Teachers' Professional Learning*, Perspektiven der Mathematikdidaktik, https://doi.org/10.1007/978-3-658-37236-1_1

causality and the reciprocal influences of these subsystems. Such a conceptu-alisation implies various interactions that contribute to teacher learning. This study, which captures the perspectives of individual teachers, focuses on internal interactions in the participating teachers' individual learning orientation system—interactions among teacher experiences, beliefs, knowledge and practices—and how the individual system interacts with the school system, which reveals both cognitive and situated perspectives on learning theory.

This study investigates teacher learning from the cognitive perspective to focus on individuals' construction of knowledge. Analyses that focus on individual teachers are usually examples of the cognitive perspective of learning (Ander-son, Reder, & Simon, 1997; Cobb & Bowers, 1999), as they tend to emphasise teachers' experiences, beliefs, knowledge and practices. Richardson (1996, 2003) found that preservice teachers' beliefs were shaped by their experiences. Both teacher beliefs and knowledge are important for mathematics teachers' teach-ing—teachers' pedagogical beliefs guide their classroom teaching, but the ways in which they enact those beliefs are likely constrained by the knowledge they possess (Fennema & Franke, 1992; Zhang & Wong, 2014). Teachers' beliefs may also be enacted when they perform related tasks, particularly when they experi-ence success in doing so (Bandura, 1997; Buehl & Beck, 2014). Researchers have argued that the dissonance between various aspects—including teacher experience, beliefs, knowledge and practice—may result in teacher learning (Woolfolk Hoy, Hoy, & Davis, 2009). According to Opfer and Pedder (2011), the interactions and intersections among these factors constitute an individual-level orientation towards teacher learning. However, much of the extant literature focuses exclusively on individual aspects of these interactions; a more holistic investigation of all aspects of the teacher system and the interactions among those aspects is required to fully assess teacher professional learning.

The study also adopts a situated perspective that emphasises the interac-tions between individuals and their environment. The situated perspective focuses on the teaching environment and emphasises a school-level orientation towards teacher learning. Opfer and Pedder (2011) argued that a school constructs organ-isational learning via the interactions among its beliefs, knowledge and practices. Teachers' practice and learning are simultaneously influenced by the school's systems and particularly by the norms that inform the context in which teach-ers' professional learning is situated (Tschannen-Moran, Salloum, & Goddard, 2014). Inexperienced teachers are more easily impacted by their surroundings (Woolfolk Hoy & Burke-Spero, 2005); a centralised curriculum, for example, pro-vides valuable supports for teachers' practices during the crucial initial stages of their teaching careers (Haser & Star, 2009). These supports are believed to exert long-term career influences (Feiman-Nemser, 1983; (Organisation for Economic Co-operation and Development [OECD], 2005).

The study was conducted in China[1], an East Asian country in which mathematics education has already attracted researchers' attention owing to East Asian students' consistent outstanding achievements in large-scale international assessments, such as TIMSS (Trends in International Mathematics and Science Study) and PISA (Programme for International Student Assessment) (Mullis, Martin, Foy, & Arora, 2012; OECD, 2012). Moreover, while China and nearby countries share a culture that is rooted in Confucianism and its values and paradigms are recognised as an essential factor impacting the learning and teaching of mathematics (Leung, 2001). While its importance has been widely recognised (e.g. An, Kulm, Wu, Ma, & Wang, 2006), research into how teachers in these regions practice teaching and engage in professional pedagogical learning remains scarce (Fan, Miao, & Mok, 2014).

Shanghai was selected as the context for this study, as it is typical of locations that are highly influenced by Confucian culture and is an attractive research site in the field of mathematics education, owing to the high rating it obtained in PISA. Moreover, the Shanghai setting provides the researcher with convenient access to the necessary information. For a comprehensive understanding of teacher professional learning from both the cognitive and situated perspectives, a case study approach was employed (see Chapter 3). Three novice mathematics teachers working in different Shanghai schools were recruited to report their interactions with their situated contexts. The construction of their teacher learning in all its individual aspects (experiences, beliefs, knowledge and practices) was considered to present a holistic picture of their teaching practice and learning.

1.2   Aims of the Study

Through a two-year investigation, this study explored how novice mathematics teachers in Shanghai experienced professional learning in schools to develop and verify the theory that teacher professional learning is complex and dynamic and to appraise the practice of teacher education based on the implications obtained from the three teachers' teaching practice. To realise these aims, two general research questions, which are further divided into several sub-questions in Chapter 3, were addressed:

---

[1] This study focuses primarily on mathematics teaching in mainland China.

1. What learning outcomes could the novice mathematics teachers obtain during the observed two years of teaching, particularly in terms of teacher beliefs, teacher knowledge and teaching practice?
2. How was professional learning implemented to achieve the learning outcomes?

1.3   Overview of Research Methodology

To achieve the above objectives, the study adopted a case study approach to three novice mathematics teachers in different upper secondary schools in Shanghai, each of whom had commenced their teaching practice in September 2013. The case study approach allowed for a two-year longitudinal exploration of the initial stages of the teachers' teaching and the collection of rich data from multiple sources. It also permitted an in-depth investigation of the complex teacher professional learning process as well as the specific contexts in which the participants were situated and embedded.

The investigation comprised four rounds of data collection, spread over four semesters in a two-year period. The instruments used in each round included classroom observation, interviews, document reviews and field notes. Classroom observations were conducted for three or four consecutive lessons for each teacher in each semester, amounting to a total of thirty-eight lessons. All lessons were videotaped to capture the teachers' classroom teaching styles and to allow the researcher to explore the beliefs and knowledge they enacted in their teaching. For each lesson, a pre-observation interview was conducted to reveal how the teacher had prepared the lesson, while post-observation interviews allowed the teachers to reflect on their teaching. Interviews were also conducted separately during the first and fourth rounds of data collection to reveal and investigate any changes in the teachers' beliefs and knowledge. To triangulate the researcher's interpretation of the teachers' practice and learning, interviews were also conducted at the study period's conclusion, in which the novice teachers reflected on their professional learning over the two years and their mentors reflected on their pedagogical beliefs and their interactions with their mentees. Teaching-related documents were also collected and field notes were taken throughout the investigation.

All classroom observations and interviews were transcribed verbatim. To accurately represent the learning process, data were collected at four time points to present the three stages of the two-year period: the beginning, medium and final stages. The analysis during the beginning stage focused on the holistic features of the teachers' teaching practice as originally implemented and the beliefs and knowledge they brought into teaching, based on what they stated in the interviews and how they behaved in classroom teaching. In the second stage, the analysis focused on the main features of their teaching practices and on the beliefs and knowledge reflected in their teaching to identify any changes that may have occurred as a result of their professional learning. At the final stage, similar analyses explored which aspects of the teachers' practices, beliefs and knowledge remained consistent or changed as a result of their professional training. Through the above analyses, a holistic description of the three teachers' teaching over the two-year period was developed (see Chapters 4 to 6). The findings were then summarised and compared across the various stages and teachers using inductive analysis, and discussions based on related literature resulted in the extraction of conclusions from an iterative review of the data and findings.

1.4   Significance of the Study

The study's significance is discussed in detail in Chapter 8 from both the theoretical and practical perspectives. From a theoretical perspective, it offers insights into case-study research methodology and longitudinal research design; verification and enrichment of the theory of teacher professional learning; and comprehensive understanding of mathematics teacher education and the learning and teaching of mathematics in the context of China. From a practical perspective, the study is expected to highlight implications for preservice and in-service teacher education in similar contexts, with a particular focus on the teaching and learning of novice mathematics teachers in light of individual teacher differences and environmental influences.

1.5   Overview of the Book

This chapter has briefly introduced the study's rationale, aims, research methodology and significance as well as the complex and dynamic conceptualisation of the cognitive and situated perspectives on learning. Chapter 2 reviews the relevant literature, outlines the theory underlying the study and emphasises the influences of both the individual and school levels on novice teachers' professional learning. It focuses on mathematics teachers' beliefs, knowledge and practices and the impact of the school as well as the relationships among these aspects. Chapter 3 details the ways in which these aspects are examined, establishes the study's research methodology and justifies its design within the qualitative paradigm.

The subsequent three chapters present the individual differences and similarities that came to light in the three teachers' professional learning during the study period. Chapter 4 presents the case of Doris, who trained as a mathematics teacher in a four-year bachelor's programme, a one-year teaching practicum and a three-year master's programme prior to beginning work in School A. During the two-year teaching period, she focused on learning to teach school mathematics in ways that were consistent with the collective ideas of other teachers in the same environment while making efforts to promote students' interest and mathematical thinking by integrating mathematical history and culture into her teaching. Chapter 5 examines Jerry's case: he had undergone teaching-related training during his bachelor's programme and learned pure mathematics in his master's programme. During the two-year study period, his teaching emphasised students' performance in the college entrance examination (*Gaokao*), as demanded by School B's environment. Chapter 6 focuses on Tommy, who did not receive any teaching-related training and instead focused on learning advanced mathematics before joining School C as a teacher. He paid particular attention to learning how to teach mathematics in a school context and changed his existing pedagogical beliefs accordingly.

Chapter 7 further investigates the mentorship that the three novice teachers experienced during the two-year's professional learning. Based on the mentees' perspectives, it summarises the various mentorships in China and discusses the opportunities and constraints that the different mentorships presented to the novice teachers.

Chapter 8 synthesises and discusses the main findings from the three case study chapters to respond to the study's two research questions. It also discusses the study's significance and limitations and finally lists several suggestions for further research.

Literature Review and Theoretical Framework

2

This chapter contextualises the study in relation to the theoretical literature to explain the theoretical perspectives on teacher professional learning applied throughout. Both the cognitive and situated perspectives of learning theory are considered. Following the cognitive perspective, the chapter focuses mainly on issues relating to the individual-level approach to teacher learning, including the features of individual teachers and the central elements involved in their learning. These elements—teacher beliefs, knowledge, and teaching practice—and the interactions among them contribute to a multi-faceted construct for teacher learning. Informed by this construct, we carefully review the substantial body of research on these elements and their interactions with particular focus on operational definitions, frameworks for their description, and analytical approaches to them. Additionally, studies on the school environment's influence on teacher learning from the situated perspective are also discussed. Based on the literature presented, the study's general conceptual framework is proposed at the end of the chapter.

2.1 Teacher Professional Learning

In seeking ways of improving schools and the quality of students' learning, researchers have paid considerable attention to teachers' professional learning (Opfer & Pedder, 2011; Sowder, 2007). Accordingly, multiple theories and studies have emerged in this field. This section first clearly defines the concept of and interchangeable terms for teachers' professional learning and then reviews the concept's theoretical background before finally presenting the specific theory adopted in this study.

© The Author(s), under exclusive license to Springer Fachmedien Wiesbaden GmbH, part of Springer Nature 2022
X. Lu, *Novice Mathematics Teachers' Professional Learning*, Perspektiven der Mathematikdidaktik, https://doi.org/10.1007/978-3-658-37236-1_2

2.1.1   A Delimitation of Teacher Professional Learning

Several similar terms relate to and are sometimes used interchangeably with the term 'teacher professional learning', including 'teacher professional development', 'teacher change', and 'teacher growth'. According to Clarke and Hollingsworth (1994, p. 154), teachers' professional development denotes in-service programmes and is a generic term encompassing a particular research stance. The use of different terms usually points to the researcher's specific perspective; teacher change, for example, refers to a process, an observable phenomenon, or a set of behaviours, while teacher growth encompasses both a change process and an intention to invoke learning. The term adopted in this study—teacher professional learning (TPL)—refers to teachers' learning how to learn and how to transform their knowledge into practice for the benefit of their students' growth (Avalos, 2011).

2.1.2   Theoretical Background of TPL: The Two Perspectives on Learning Theory

The term 'teacher professional learning' appears consistent with the 'teacher growth' concept proposed by Clarke and Hollingsworth (2002) in that it 'must conform to some coherent theory of learning' (p. 955). Two perspectives on learning theory exist—cognitive and situated—which adopt the individual and the social collectives, respectively, as the primary units of analysis (Cobb & Bowers, 1999).

Cobb and Bowers (1999), based on Sfard's (1998) focus on core metaphors, suggested that a central metaphor for the cognitive perspective is to consider learning as an entity acquired in one task setting that can also be conveyed in other settings; a primary metaphor for the situated perspective is to regard learning as an activity in which individuals are situated in circumstances that involve abundant social affairs. Originally, researchers tended to situate the two perspectives on opposite sides by detailing their distinctions in their underlying assumptions and methodologies (Anderson, Reder, & Simon, 1996; Anderson, Reder, & Simon, 1997; Greeno, 1997). However, Cobb and Bowers (1999) observed that the two perspectives are not easily distinguished from each other; for instance, they found that Anderson et al. (1997)—despite claiming their unit of analysis to be the group or local community—actually focused on individuals' cognitive activity.

The two perspectives on learning theory are also used to interpret teachers' learning, depending on whether researchers view learning as the development of knowledge or of practice (Clarke & Hollingsworth, 2002). More recently, Postholm (2012) connected learning theory to a cognitivist/constructivist paradigm. From a cognitivist perspective, she asserted that learning happens when one is taught or is mentally situated in different ways, while the constructivist view perceives teachers' knowledge as the construction of meanings and understandings within social interaction. Thus, in Postholm's view, teachers themselves are in the foreground of the learning process, constructing knowledge and actively facilitating learning through mediated acts in their surroundings. In this sense, both perspectives are involved in teachers' professional learning.

2.1.3 Adopting a Theory that Considers the Complexity and Dynamics of TPL

Researchers recognise the complexity and dynamics of TPL from both the cognitive and situated perspectives (Opfer & Pedder, 2011; Avalos, 2011). It is widely accepted that simplified conceptualisations of TPL may lead to ineffective professional learning activities (e.g., Borko, 2004; Timperley & Alton-Lee, 2008). However, the extant literature focuses largely on specific activities or programmes for teachers' learning rather than on the complex teaching and learning environment; this is also true of mathematics teachers' learning. Goldsmith, Doerr and Lewis (2014) reviewed 100 studies in the field and found them to be lacking in consistent descriptions of the elements of TPL, common frameworks for describing learning, and common approaches to assessing learning. In particular, the existing literature focuses excessively on whether a programme is effective with respect to teachers' practice and students' learning while investing too little effort in understanding teachers' learning (Goldsmith et al., 2014).

To obtain a comprehensive understanding of novice teachers' learning in the early stages of their teaching careers, this study follows the conceptualisation advocated by Opfer and Pedder (2011), which involves methodological practices focused on explanatory causality and the reciprocal influences of various elements involved in three systems centred on the teacher, the school, and the learning activity. With a particular emphasis on the cognitive perspective of teacher learning, this study examines teachers' experiences, beliefs, knowledge, and teaching practices—all of which are considered elements of the teacher system while the interactions among them are considered to be the orientation of teacher learning (Opfer & Pedder, 2011). The study also considers—from the situated perspective

on learning—how surroundings affect teachers' learning in the particular context of China. According to Opfer and Pedder (2011), elements in the school
system—specifically, collective orientations and beliefs about learning, collective
practices or norms of practice, and the collective capacity to realise shared learning goals—are similar to those in the teacher system, and the interactions between
these elements and the teacher system promote teachers' learning. The sections
that follow will present the literature to demonstrate how the essential elements of
this study will be investigated, including related theories or frameworks, previous
findings, and assessment methods.

2.2   Novice Teachers

From the cognitive perspective, the individual is the principal theoretical focus
(Anderson et al., 1997). As Avalos (2011) stated, numerous studies have focused
on novice teachers' learning as this has been well recognised as a particularly
complex stage worldwide (OECD, 2005). This section first presents the features
of this stage and the psychological experiences of novice teachers and then illustrates the unique environment (mentoring and induction) that particularly supports
their professional learning.

2.2.1 The Stage of Expertise

The beginning stage of teachers' teaching careers, according to the stage theories
of expertise, usually occurs in their first three years of practice (e.g., Huberman,
1993; Berliner, 1988). Based on the skill acquisition process outlined in Dreyfus and Dreyfus' (1987) stage model, Berliner (1988) presented a general theory
of teaching skills development and illustrated the features of novice teachers by
comparing existing findings on two extreme developmental stages (novice and
expert). Berliner claimed that teachers' behaviour in their initial year is rational
and relatively inflexible; at this stage, experiences gained from teaching practice
are more important to them than verbal information, and they are accustomed to
following rules and procedures established by others. By their second or third
years, novice teachers have developed sufficient strategic knowledge to contextu-
alize their behaviour; however, because of their strict adherence to rules, novice
teachers may lack responsibility for their actions to some extent.

Lian (2008) surveyed 3000 teachers in China (including novice, proficient, and expert teachers) to study their 'mental experiences', as they were termed by Lian himself, and identified two sub-stages in novice teachers' learning (Lian, 2008); during the first sub-stage (their first and sometimes second years), they have strong external motivation to succeed and pursue basic teaching knowledge, while by the second sub-stage (their second and/or third years), they have mastered basic teaching skills and realise the complexity of teaching improvement. Novice teachers might obtain the necessary teaching experience and become proficient over three to five years of teaching practice; however, some might also doubt their own competence during this stage, and may become increasingly unenthusiastic and unmotivated with respect to the future.

The developmental processes involved in expertise acquisition are not always clear-cut, and expertise is highly contextualised (Berliner, 1988). However, stage theory is not that it establishes boundaries between the stages but that the stages can help us to make sense of our thinking about teacher expertise. The stage theory of teacher expertise contributes to our understanding of the extent to which novice teachers learn through their acquisition of teaching skills and their 'mental experiences'; however, the duration of each stage of the expertise development process is not confined and may differ significantly across different contexts. Thus, in-depth research on novice teachers in specific contexts is required, and teachers' cognitive learning processes and teachers' situations should both be focal points for research.

2.2.2   Features of Novice Teachers in China

Studying teachers' 'mental experiences', Lian (2008) found that, compared with experienced teachers, novice teachers in China (1) focus more on lesson preparation before class; (2) are more lively, enthusiastic, and outgoing; (3) focus more on their performance; (4) have lower career commitment but higher professional burnout rates; and (5) can feel external support and satisfaction.

First, owing to the lack of coordination ability in classroom teaching, novice teachers emphasise pre-class preparation, even to the point of preparing the precise words that they will use in class, and usually rely on these preparations when implementing classroom instructions. Second, novice teachers are anxious to pursue self-actualisation and gain recognition in their new environment. Third, they focus more on external assessment of their teaching. Fourth, they must develop long-term teaching practices to realise good occupational psychology. Finally, novice teachers have more positive emotions around teaching, as they perceived greater support from school leaders and the teacher community (Lian, 2008).

Three of the above findings mention environmental influences, while the others focus on teachers' own cognition. Lian's 'mental experiences' involve interactions between novice teacher beliefs, knowledge, and teaching practices in situated contexts. These findings provide us with an overall picture of early career Chinese teachers' unique 'mental experiences' and suggest that the beginning stage may play a pivotal role in the teachers' career trajectory. However, as it employed a quantitative research method, Lian's (2008) study did not show in detail how the 'mental experiences' were constructed and changed.

2.2.3 Mentoring and Induction

The beginning stage of teaching is complex and important (Hebert & Worthy, 2001; OECD, 2005), and researchers have conducted various studies to make comparisons between novice and experienced teachers and investigate mentoring and induction (Avalos, 2011). International literature provides various perspectives on the purposes of mentoring. By examining the mentoring process, researchers provide, for example, suggestions for policy-makers (e.g., Hobson, Achby, Malderez, & Tomlinson, 2009), information on the role and nature of mentors (e.g., Harrison, Dymoke, & Pell, 2006; Hennissen, Crasborn, Brouwer, Korthagen, & Bergen, 2010) and on mentees' needs (e.g., Fantilli & McDougall 2009) as well as perspectives on the school socialisation of novice teachers (Kelchtermans & Ballet, 2002).

Although mentoring and induction programmes have gained considerable momentum in recent years, no definitive evidence for the value of mentoring has been offered (Ingersoll & Kralik, 2004). Mentoring activities for early career teachers have been proposed internationally as a possible means of achieving their professional learning and have a long tradition in East Asia. Mentoring activities aim to assist novice teachers in situating themselves within the school community and help them to comply with the requirements of their new position in the induction phase, so mentoring is usually considered supportive (Kemmis, Heikkinen, Fransson, Aspfors, & Edwards-Groves, 2014).

Orland-Barak (2014) provided a systematic review of studies on mentoring across various contexts and identified three topics:

1) Mentors' performance and behaviours;
2) Mentors' reasoning, beliefs and identity formation; and

3) The place of culture, context and discourse.

Of the 39 articles reviewed, only one covered an East Asian context in which mentoring has long existed as an important school-based training means of familiarising teachers with the teaching environment within a short time (e.g., Shi, 2002).

Mentoring is shaped by interactions between mentors and mentees in professional settings; thus, the dialogues between mentors and mentees are usually analysed for the purpose of studying mentoring. Hennissen, Crasborn, Brouwer, Korthagen, and Bergen (2008) identified five aspects of mentoring dialogues based on an analysis of 26 publications:

1) Content and topics, referring to the content of the dialogues (e.g., instructional and organisational issues);
2) Style and supervisory skills, focusing on the specific supervision skills of the mentor, directive or non-directive style;
3) Mentor's input, referring to the person who initiates dialogues and the level of participation, either active or reactive mentors;
4) Times aspects, referring to the duration of the mentoring dialogues and mentors' speaking time during the dialogues;
5) Phases, focusing on topics concerning the stage of the dialogue and the differences between trained and untrained mentors.

Based on these five aspects, a two-dimensional model was developed to evaluate mentors' roles in dialogues (Hennissen et al., 2008, p. 177). Four mentor roles were constructed: initiator—introduces topics and uses non-directive skills with short speaking time; imperator—introduces topics and uses directive skills with long speaking time; advisor—does not introduce topics and uses directive skills with long speaking time; and encourager—does not introduce topics and uses non-directive skills with short speaking time. This model by Hennissen et al. (2008) investigates mentoring through various perspectives; however, it does not consider the mentee's input. Moreover, since it was developed based on Western contexts, it may miss the consideration of East Asian contexts.

Mainland China's teaching and research system incorporates a comprehensive framework for teacher mentoring (Salleh & Tan, 2013). It is believed that mentoring helps to compensate for a lack of pre- or in-service programmes, particularly in practicum (Mao & Yue, 2011) and may facilitate novice teachers' long-term professional training, which is what they actually need (Gu, 2006). All novice teachers in Shanghai report that they are required to participate in mentoring programmes during the early stages of their careers (OECD, 2015).

However, with the development and reform of teacher education, traditional one-to-one mentoring has been criticised as overemphasising the one-way transmission of information, thus contributing to mentees' lack of knowledge and narrow thinking (Yan & Duan, 2009). In her investigation of more than 300 novice teachers in Shanghai from 2008 to 2009, Wang (2009) found that not all mentoring activities were effective, particularly when the novice teachers recognised the importance of discussion with more experienced teachers and not only their mentors; they felt as though they were betraying their mentors by discussing with other teachers. Mentoring appears to play a key role in novice mathematics teachers' professional learning in terms of both supports and constraints.

Considering the TPL dynamic and the interactions between the individual teacher and his/her situated teaching environment, Li (2011) pointed out that the positive aspects of mentoring should be drawn on to help new teachers develop their signature scientific teaching features and styles rather than just absorb experienced teachers' knowledge of teaching skills and methods. With this in mind, local researchers have paid attention to the development of new forms of mentoring or the extension of mentoring, such as from one-to-one mentoring to group mentoring (Yan & Duan, 2009), team-work growth (Wang, 2009), or the development of professional learning communities (Song, 2012). In this sense, mentoring is not constrained to a one-to-one format. As a school-based approach to teacher training, it is embedded in and probably impacted by the school environment. Thus, in this study, mentoring was regarded as an important aspect of the school-level approach to learning.

2.3 Individual-Level Orientation to Learning

This section reviews literature on the elements of the teacher system, including teacher beliefs, knowledge and practice.

2.3.1 Teacher Beliefs and Knowledge

Novice teachers develop or obtain their professional beliefs and knowledge from their learning- or teaching-related experiences (Novak & Knowles, 1992; Richardson, 1996; Richardson, 2003). Both beliefs and knowledge are recognised as important aspects of teachers' practices and teacher change.

Teacher beliefs and knowledge are important components in the study of the TPL process. Researchers have long agreed that beliefs are the best indicators of the decisions that individuals will make throughout their lives (e.g., Bandura, 1986; Dewey, 1933); thus, teacher beliefs determine their teaching activities (Pajares, 1992; Thompson, 1992), and teacher behaviours are best understood by focusing on their beliefs (e.g., Clark, 1988; Cole, 1989). Knowledge is used in all activities and walks of life, concerns different topics, and enables various actions (Bandura, 1986); the placement of knowledgeable teachers in each classroom is essential to students' mathematical knowledge (Sowder, 2007).

2.3.1.1 Definition and Meaning

According to Pajares (1992), the term 'beliefs' is often used as synonym for attitudes, values, judgements, axioms, opinions, ideology, perceptions, conceptions, conceptual systems, preconceptions, dispositions, implicit theories, explicit theories, personal theories, perspectives, repertories of understanding, and social strategy; researchers attempt to distinguish between beliefs and other terms depending on the purpose of their studies. For example, Bishop, Seah, and Chin (2003) argued that beliefs are more contextual than values, since beliefs are used to evaluate and make judgements, while values are associated with individual desirability. However, many researchers perceive more similarities than differences between values and beliefs. Rokeach (1973) viewed values as enduring beliefs, Clarkson and Bishop (1999) thought of values as beliefs in action, and Raths, Harmin and Simon (1987) argued that beliefs constitute values (cited by Philipp, 2007). The terms 'beliefs' and 'values' are often used interchangeably and will be used interchangeably in this study.

Another controversy surrounding the meaning of beliefs is the distinction between beliefs and knowledge (Pajares, 1992). Furinghetti and Pehkonen (2002) argued that knowledge consists of two parts—objective knowledge and subjective knowledge—and that beliefs are a form of subjective knowledge; most researchers, however, distinguish between the two. Some assert, for example, that knowledge must be associated with truth (Philipp, 2007) and that a belief cannot be called knowledge unless it is true; however, this distinction is limited, since truth is difficult to define. Thompson (1992) argued that beliefs and knowledge have different degrees of inter-subjective consensus, different requirements for acceptance, and relate to different aspects of learning—specifically, knowledge is usually related to truth and certainty, while beliefs are often related to doubts and disputes. Generally, the breadth of the definition of knowledge determines whether beliefs are included; in this study, a narrow definition of knowledge is adopted.

The difficulty in distinguishing between beliefs and knowledge may stem from their influences on teaching. It is not easy to specify how beliefs and knowledge work in practice (Cooney & Wilson, 1993). Ernest (1989) held that beliefs play a regulatory role between knowledge and behaviour. Fennema and Franke (1992) asserted that both beliefs and knowledge interactively influence teaching, while Blömeke, Felbrich, Müller, Kaiser, and Lehmann (2008) considered them to be components of teachers' professional competence (cited by Zhang & Wong, 2014). This study asserts that interactions exist among teacher beliefs, knowledge, and teaching practice.

Despite these interactions, mathematics teachers' professional beliefs and knowledge are differentiated in this study to allow multiple perspectives on and clear descriptions of teachers' teaching and their learning for teaching. The study follows Raymond's (1997) definition of beliefs as 'personal judgment about mathematics formulated from experiences in mathematics, including beliefs about the nature of mathematics, learning mathematics, and teaching mathematics' (pp 552). These three facets of mathematics teacher beliefs have been widely discussed and used to investigate mathematics teacher beliefs (see 2.3.1.2). Inconsistencies have also been identified between teachers' stated teaching beliefs and their actual teaching practices (e.g. Cross, 2009; Cross, 2015; Raymond, 1997), and researchers are increasingly focused on the complexity of classroom teaching (Skott, 2009). Wong, Ding, and Zhang (2016) argued that the complex context of the classroom greatly influences teacher beliefs. To obtain a comprehensive understanding of the beliefs held by novice mathematics teachers, this study examines both their stated beliefs and enacted beliefs (i.e., actual classroom teaching practices).

For mathematics teachers, professional knowledge is a set of knowledge specific to the teaching of mathematics. Given that knowledge is an important ingredient in teaching, it is necessary to know what factors influence knowledge construction and how they interact with each other (Petrou & Goulding, 2011). However, in the absence of a single, widely accepted framework for evaluating teachers' mathematical knowledge (Tirosh & Even, 2007), this study adopts two analytical frameworks that focus on teaching practice (see 2.3.1.4).

2.3.1.2 Framework to Assess Teacher Beliefs

As mentioned above, studies of mathematics teacher beliefs usually focus on teachers' perspectives on the nature of mathematics, their teaching, and their learning of mathematics, since these contribute to people's understandings and interpretations of teachers' teaching (Philipp, 2007; Thompson, 1992).

Beliefs about the nature of mathematics are considered the dominant aspect and thus are the focus of most analyses of mathematics teacher beliefs (Chapman, 2002). Ernest (1989) argued that teacher's views about the nature of mathematics implicate their personal philosophies of mathematics and distinguished three philosophies (instrumentalist, Platonist, and problem-solving) that he speculated could be arrayed in a hierarchy running from instrumentalist (the lowest level) to problem-solving (the highest). Ernest (1989) also discussed the corresponding teacher beliefs about teaching and learning to illustrate these beliefs' impact on teaching practice (see Table 2.1).

Table 2.1 Ernest's (1989) view of mathematics teacher beliefs

Teacher beliefs about	Lowest level	Intermediate level	Highest level
The nature of mathematics	Instrumentalist view: Maths as a set of unrelated but utilitarian rules and facts	Platonist view: Maths as a static but unified body of certain knowledge	Problem-solving view: Maths as a process of enquiry and coming to know, not a finished product, for its results remain open to revision
Teaching, Teachers' role	Instructor: skills mastery with correct performance	Explainer: conceptual understanding with unified knowledge	Facilitator: confident problem posing and solving
Use of curriculum materials	The strict following of a text or scheme	Modification of the textbook approach, enriched with additional problems and activities	Teacher or school construction of the mathematics curriculum
Learning	Learning as passive reception of knowledge	Learning as the reception of knowledge	Learning as the active construction of understanding, even as autonomous problem posing and solving

Furthermore, Van Zoest, Jones, and Thornton (1994, p. 2) reviewed a study by Kuhs and Ball (1986), in which four dominant and distinctive views of or approaches to mathematics teaching were provided: learner-focused; content-focused with an emphasis on conceptual understanding; content-focused with an emphasis on performance; and classroom-focused (i.e., teaching based on research knowledge about effective classrooms). They used three of the four views (learner-interaction, content-understanding, and content-performance) to build a framework for assessing pre-service teacher beliefs about mathematics teaching; at one extreme, the learner-interaction view posits a socio-constructivist orientation of mathematics teaching, while the content-performance view reflects a performance-driven orientation at the other; the content-understanding view stands at the intermediate point between the other two views. Based on existing literature (Ernest, 1989; Van Zoest et al., 1994), Beswick (2005) summarised the theoretically consistent relationships among mathematics teacher beliefs about the nature of mathematics, mathematics teaching, and mathematics learning and emphasised the logical interrelationships and groundings in theories of the three philosophies of mathematics (as psychological systems of beliefs).

Similarly, Thompson (1991) proposed a three-level framework of mathematics teacher beliefs, characterised by five conceptual aspects (p. 9): (1) what mathematics is, (2) what it means to learn mathematics, (3) what one teaches when teaching mathematics, (4) what the roles of the teacher and students should be, and, (5) what constitutes evidence of student knowledge and criteria for judging correctness, accuracy, or acceptability of mathematical results and conclusions. Most views can be categorised according to the three aspects of teacher beliefs (i.e., about the nature of mathematics, mathematics teaching, and learning); in particular, Thompson stated different conceptions of problem-solving corresponding to the three levels.

Raymond (1997) categorised teacher beliefs about the nature of mathematics, mathematics teaching, and learning on a five-level scale, ranging from traditional to non-traditional, based on Ernest's (1989) categories. On Raymond's scale, traditional corresponds to instrumentalist; an even mix of traditional and non-traditional corresponds to Platonist; non-traditional corresponds to problem-solving; primarily traditional lies between traditional and the even mixture (i.e., closer to instrumentalist than to Platonist); and primarily non-traditional lies between non-traditional and the even mix (i.e., closer to problem-solving than to Platonist).

From the above literature, a hierarchical theoretical framework for mathematics teacher beliefs emerges (see Table 2.2) in which three components are involved: beliefs about the nature of mathematics; beliefs about mathematics teaching; and beliefs about mathematics learning. Of these, beliefs about the nature of mathematics dominate and are theoretically consistent with the other two (Beswick, 2005). The beliefs generally range from viewing mathematics as a static, procedure-driven body of facts and formulas to considering mathematics a dynamic domain of knowledge based on sense-making and pattern-seeking, although there are several intermediary categories detailed in the literature (Cross, 2009; Zhang & Wong, 2014). For convenience, unified category labels are used in this study: traditional, mixed traditional and non-traditional, and non-traditional. Other facets, including beliefs about mathematics teaching and learning and problem-solving, exhibit distinct characteristics at different levels (as summarised in Table 2.2), based on the research of Beswick (2005), Ernest (1989), Kuhs and Ball (1986), Thompson (1991), and Van Zoest et al. (1994).

In Table 2.2, beliefs about the nature of mathematics are characterised in terms of teachers' views about mathematics and what mathematics is; beliefs about mathematics teaching, including what mathematics teaching is, the use of curriculum materials, the teacher's roles, intended outcomes and judging correctness; beliefs about mathematics learning, particularly what is learning and what are the roles of the student; and beliefs about problem-solving, focusing on the purpose thereof. This detailed summary is used as the framework for the researcher to clarify questions in interviews and to analyse participants' personal beliefs and beliefs reflected in teaching activities.

Table 2.2 Summary of the three levels of mathematics teacher belief

	About Mathematics	About Mathematics Teaching	About Mathematics Learning	About problem-solving
Traditional category (Instrumental-ist)	Mathematics is a set of unrelated but utilitarian rules and facts; Common uses of arithmetic skills in daily situation.	*Content-focused with an emphasis on performance* Teaching emphasises student performance and mastery of mathematical rules and procedures and is a way to demonstrate well-established procedures; The teacher follows a text or scheme strictly; The teacher is an instructor or demonstrator; The intended outcome of teaching is skill mastery with correct performance; Authority for correctness or accuracy lies in the teacher or the book.	Learning is skill mastery, passive reception of knowledge; The learner is an imitator.	Solving problems is getting answers by using prescribed procedures.
Mixed of traditional and non-traditional (Platonist)	Mathematics is a static but unified body of certain knowledge; Except for rules and procedures, there are concepts and principles 'behind the rules'.	*Content-focused with an emphasis on understanding* Teaching is driven by the mathematical content itself but emphasises conceptual understanding and includes the usage of instructional representations and manipulatives; The teacher modifies the textbook approach and enriches it with additional problems and activities; The teacher is an explainer; The intended outcome of teaching is conceptual understanding with unified knowledge; Authority for correctness or accuracy lies with experts.	Learning is the reception of knowledge but the active construction of understanding; Learning mathematics is funny; The learner is somewhat broadened to include some understandings.	Problem-solving is important in mathematics curriculum, but is a separate curricular strand to be taught separately.

(continued)

Table 2.2 (continued)

	About Mathematics	About Mathematics Teaching	About Mathematics Learning	About problem-solving
Non-traditional (problem-solving)	Mathematics is a process of inquiry and coming to know, not a finished product, for its results remain open to revision.	*Learner-focused* Teaching focuses on the learner's personal construction of mathematical knowledge and develops students' reasoning; Teacher or school construction of the mathematics curriculum; The teacher is a facilitator, steering students' thinking in mathematically productive ways; The intended outcome of teaching is confident problem-posing and -solving; Students' judgements.	Learning is the active construction of understanding, autonomous exploration of own interests; and the learner must engage in mathematics inquiry when making sense of mathematical ideas.	The process of doing mathematics is the way to grow understanding.

2.3.1.3 Ways to Study Mathematics Teacher Beliefs

Teacher beliefs are best examined based on inferences drawn from what they say or do (Philipp, 2007; Pajares, 1992). As such, two typical approaches to studies on teacher beliefs exist—assessment instruments and case study methodology. For large groups of teachers, quantitative methods such as Likert scales are often used to measure their beliefs. For example, the Teacher Education and Development Study in Mathematics (TEDS-M), a large-scale teacher professional development study, used several belief scales formulated from the literature to measure student teacher' beliefs; however, researchers have questioned the validity of Likert-scale surveys (e.g., Ambrose, Clement, Philipp, & Chauvt, 2004), claiming that the item's wording may not be sufficiently clear to prevent misinterpretation. To address this shortcoming, another approach—case study—is typically applied.

Case studies allow researchers to describe teacher beliefs based on rich data sets derived from classroom observations, interviews, surveys, concept mappings, and other sources. Considering the complex environments in which teachers are situated, a case study is usually conducted over a period of time and often adopts the triangulation of data. For instance, Raymond (1997) investigated the inconsistencies between teacher's mathematics beliefs and teaching practice by conducting a case study of six novice elementary school teachers. A variety of data collection methods, including questionnaires, interviews, classroom observations, document reviews, and concept-mapping activities, were used during the teachers' first full-time teaching year. Qualitative research methods are expensive and time-consuming, and can accommodate only a small number of participants. However, the deep descriptions and analyses they yield support theory building and enable researchers to rise to the challenge (proposed by Thompson (1992)) of investigating the dialectic relationship between teachers' conceptions, including their beliefs, knowledge, and teaching practice. Thus, qualitative research methods were employed in this study to investigate novice mathematics teacher beliefs, both stated and enacted.

2.3.1.4 Frameworks for Assessing Teacher Knowledge

Two frameworks of teacher knowledge mentioned in this study stem from Shulman's categories of teacher knowledge for effective teaching. Shulman (1986, 1987) constructed four general dimensions of teacher knowledge (general pedagogical knowledge, knowledge of learners and their characteristics, knowledge of educational contexts, and knowledge of educational values and purposes) and three content dimensions of teacher knowledge (subject content knowledge, curriculum knowledge and pedagogical content knowledge). These content-specific dimensions are highly related to teaching and were emphasised by Shulman. As argued by Goulding, Rowland, and Barber (2002), these classifications are generic but can easily be applied to the mathematics discipline.

Subject content knowledge (SCK) entails knowledge of the subject and its organising structures and was divided by Shulman and Grossman (1988) into substantive knowledge (key facts in the discipline) and syntactic knowledge (reasoning within the discipline). Pedagogical content knowledge (PCK) contributes to teachers' transformation of their own knowledge to their students by using analogies, illustrations, examples, explanations and demonstrations. Curricular knowledge (CK) is contained in designed programmes for teaching particular subjects or topics at a specific level and includes lateral and vertical curriculum knowledge (lateral knowledge is that which relates to knowledge in other subjects; vertical knowledge is knowledge within the same subject, but presented at different times).

Shulman's study of teacher knowledge was ground-breaking and far-reaching. However, teacher knowledge often develops in communications with students on subject matter, which is a dynamic aspect of knowledge that Shulman does not seem to acknowledge (Fennema & Franke, 1992). The distinction between SMK and PCK is also blurred (Ball, Thames, & Phelps, 2008). For example, teachers who have various representations of mathematical concepts might draw upon these in a variety of teaching plans and interactions with students; in such cases, are these representations SMK or PCK? Perhaps, as Hashweh (2005) suggests, Shulman's conceptualisation ignores the interactions between different knowledge categories. Based on their work on mathematics teaching, Fennema and Franke (1992) argued that teacher knowledge is interactive and dynamic in nature and that its development happens in specific contexts through interactions with the subject matter and the students. Therefore, teacher knowledge cannot be conceptualised outside of the situation in which the teaching is delivered.

In recent decades, a team at the University of Michigan has focused on a practice-based theory of content knowledge for both mathematics teaching and the mathematics used in teaching (Ball & Bass, 2003; Ball & Cohen, 1999; Ball et al., 2008). Their work builds on Shulman's theory and helps clarify distinctions between SMK and PCK (CK is involved in PCK). Their work resulted in a framework of mathematical knowledge for teaching (MKT), the various knowledge categories of which are summarised in Table 2.3. The framework's value lies in its ability to identify the relationship between mathematical teachers' knowledge and students' achievement. It also contributes to the assessment of mathematics teacher knowledge, such as through TEDS-M (Tatto, Schwille, Senk, Ingvarson, Peck, & Rowley, 2009). Nevertheless, Petrou and Goulding (2011) argued that Ball et al.'s categories—particularly their definition of SCK—are insufficiently clear and ignore the significance of teacher beliefs.

Table 2.3 The category of knowledge from MKT, summarised from Ball et al. (2008)

Category		Description
SMK	Common content knowledge	Mathematical knowledge and skill used in settings other than teaching Teachers need to know the material they teach; they must recognise when their students give wrong answers or when the textbook gives an inaccurate definition
	Horizon content knowledge	An awareness of how mathematical topics are related across the span of mathematics included in the curriculum
	Specialized content knowledge	Mathematical knowledge and skill unique to teaching It is mathematical knowledge not typically needed for purposes other than teaching In looking for patterns in student errors or in sizing up whether a nonstandard approach would work in general, as in our subtraction example, teachers must perform a kind of mathematical work that others do not. This work involves an uncanny kind of unpacking of mathematics that is not needed—or even desirable—in settings other than teaching.
PCK	Knowledge of content and students	The conceptions and preconceptions that students of different ages and backgrounds bring with them to the learning of those most frequently taught topics and lessons Knowledge that combines knowing about students and knowing about mathematics Teachers must anticipate what students are likely to think and what they will find confusing. When choosing an example, teachers need to predict what students will find interesting and motivating. When assigning a task, teachers need to anticipate what students are likely to do with it and whether they will find it easy or hard. Teachers must also be able to hear and interpret students' emerging and incomplete thinking, as expressed in the ways that pupils use language. Each of these tasks requires an interaction.

(continued)

Table 2.3 (continued)

Category		Description
	Knowledge of content and teaching	The ways of representing and formulating the subject and make it comprehensible to others Teachers sequence particular content for instruction. They choose which examples to start with and which examples to use to take students deeper into the content. Teachers evaluate the instructional advantages and disadvantages of representations used to teach a specific idea and identify what different methods and procedures afford instructionally.
	Knowledge of content and curriculum	n/a

In addition, another research team from the University of Cambridge developed a theoretical framework called Knowledge Quartet (KQ) that considers the practical applications of teacher knowledge and which they have used to observe and analyse novice mathematics teachers' use of SMK and PCK in classroom teaching since 2002. The KQ framework comprises four dimensions—foundation, transformation, connection, and contingency (Turner & Rowland, 2011)—in which 20 codes are used to analyse different classroom situations (summarised in Table 2.4). The foundation dimension includes SMK and teacher beliefs, while the other three align with PCK. This framework does not focus on distinctions between SMK and PCK but on how knowledge works in teachers' classroom teaching.

Table 2.4 The Knowledge Quartet framework

Knowledge Quartet	Contributory codes
Foundation	Awareness of purpose; identifying errors; overt subject knowledge; theoretical underpinning of pedagogy; use of terminology; use of textbook; reliance on procedures
Transformation	Teacher demonstration; use of instructional materials; choice of representation; choice of examples
Connection	Making connections between procedures; making connections between concepts; anticipation of complexity; decisions about sequencing; recognition of conceptual appropriateness
Contingency	Responding to children's ideas; use of opportunities; deviation from agenda; teacher insight

According to Rowland (2013), the MKT and KQ frameworks are both practice-based theories of knowledge but differ in their purposes, methods, and some outcomes. These make the two theories complementary, in that each has useful perspectives to offer to the other. This study adopts multiple perspectives from the two theories; for example, it refers to the categories of knowledge from MKT to gain a better understanding of knowledge and also follows how the KQ framework analyses classroom episodes to assess teachers' teaching.

2.3.1.5 Measuring Teacher Knowledge

Reviewing earlier scholarly research, Hill, Sleep, Lewis, and Ball (2007) found that researchers increasingly evaluated the knowledge held by mathematics teachers by studying teachers and their teaching rather than the relationship between teachers' knowledge and students' mathematics achievements. In other words, researchers started to focus on teachers' discipline-specific knowledge of content, students, and instruction required for teaching, and study in practice situations. Based on this, Hill et al. (2007) identified three main strategies for measuring professionally situated mathematical knowledge. First, observing teaching practice, directly or via video-recordings, allows researchers to carefully analyse, explain and present the mathematical knowledge that teachers use in their teaching and to assess that teaching. Second, mathematical interviews and tasks can often be used to explore teacher knowledge. These mathematical tasks differ from mathematics assessment in that researchers focus on teachers' task performance rather than on the knowledge *per se* to understand the nature and extent of teachers' knowledge. The tasks can be delivered in the form of paper and pencil tests, interviews, or a combination of the two. Third, tools can be developed along with other technologies—such as discourse analyses (e.g., Sowder, 1998) and examination of teachers' responses to video clips of teaching (e.g., Kersting, 2008)—to test the relationship between mathematical knowledge for teaching and student achievement and to track changes in teachers' mathematical knowledge.

Each method has its advantages and disadvantages. The selection of any method will depend on the specific research question and the research's scale, context, and design. In this study, teacher knowledge will be described as situated in teaching practice and informal mathematical interviews, as will be detailed in the methodology section.

2.3.2 Teaching Practice

In the field of education research, teaching practice (or teaching activity) is a broad concept that leads to research not only into activities *per se* but also into the factors that underpin teaching activities, such as teacher beliefs and knowledge. Teaching practice, in this study, refers to teaching activity. Researchers (e.g., Franke, Kazemi, & Battey, 2007) have studied teaching in the classroom context (i.e., classroom instructional practices), but this is simply one aspect of teaching activity—presentation; another important aspect in which teachers invest far more time is preparation. Burden and Byrd (2007) categorised teachers' activities into teacher planning (e.g., lesson planning, daily planning, weekly planning, unit planning, term planning, and course planning), differentiating instruction for diverse learners, selecting instructional strategies, managing instruction and the classroom, assessing and reporting student performance, grading systems, marking and reporting, and even working with others (e.g., colleagues and parents), all of which occur in the preparation and presentation parts.

To help teachers obtain effective training, researchers have characterised and structured teacher activities for teacher training and development. For instance, Borich (2000) characterised six levels of teacher training and development activities, including the structure and organisation of a teaching method or technique; classroom dialogues and examples illustrating the method or technique; practice activities and exercises; classroom observation; problem-solving activities; and decision-making dialogues. Among these, the first three are effective teaching methods while the remaining three are observation skills for effective teaching. As Borich argued, these six levels provide a sequence of learning for novice teachers. However, in a real learning situation, it makes more sense for novices to study the learning perspective.

The two aspects discussed above—lesson preparation and lesson presentation—are often used to study mathematics teaching practices. The preparation aspect mainly refers to lesson design, including content design, instructional scheduling, and teaching materials used for planning; the presentation aspect includes lesson structure (e.g., ways to construct knowledge), instructional expression (e.g., instructional strategies, language usage), materials used (e.g., technology facilities), and assessment strategies. In addition, studying teaching practice usually takes the form of classroom observation; video analysis is another popular method adopted in this study.

As participants' knowledge of teaching will be analysed separately, the present study focuses on the variety of teaching skills or strategies through which teachers apply their knowledge in practice, such as instructional strategies, both direct (e.g., presentations, demonstrations, questions, practice and drills, and guided practice and homework) and indirect (e.g., inquiry lessons, discussions, cooperative learning, and independent work) (Borich, 2000; Burden & Byrd, 2007). Teaching strategies have different characteristics in different situations. According to Coyne, Kameenui, and Carnine (2010), effective strategies for teaching mathematics include designing instruction around big ideas, designing conspicuous strategies, designing mediated scaffolding, designing primed background knowledge, designing strategic integration, and designing judicious review. These methods establish the teacher as a decision maker, with knowledge as resources and beliefs as orientations (Schoenfeld, 2010).

It is also important to consider the study of teaching practice in Chinese contexts, which may have implications for this study. In Chinese researchers' views, elementary mathematics education in China (i.e., Grades 1 to 12) has developed its own characteristics through long-term teaching practices (Fan, Miao, & Mok; 2014; Tu & Song, 2006; Yang, 1995; Zhang, 2010). A comprehensive review of related studies revealed nine aspects of Chinese mathematics teachers' teaching style: planning lessons systematically; emphasising 'two basics'; whole-class teaching and interaction; teaching with variation; teacher–student interaction and engagement; assigning and making homework frequently; using textbooks with deep understanding; structured instruction; and implementing change in light of curriculum innovations (Fan et al., 2014). In addition, Zhang (2010) mentioned that Chinese mathematics education is characterised by the introduction of new mathematical concepts; teachers implementing effective 'trying pedagogy'; interaction between the teacher and students in large classes; solving mathematics problems with the guide of variation (变式); advocating and refining 'mathematical thinking methods'; and appropriate interpretation of the mathematics education philosophy 'practice makes perfect (熟能生巧)'. All of these characteristics require high-quality teachers.

In teacher education, particularly for novice teachers, the term 'basic teaching skills (教师基本功)' is often used in pedagogy textbooks (Li, 1991; Li & Lian, 1997), teacher activities (e.g., teaching competitions), and in teacher education colleges for training student teachers, but it is less prevalent in research literature. Basic teaching skills include mathematical knowledge, skills in doing research on and coping with curriculum materials, choosing and using different instructional methods, questioning and guiding, blackboard writing, uses of modern education, and technology teaching skills in classroom instruction, among others. For instance, regional competitions for young mathematics teachers in upper secondary schools include two aspects: one involves general skills (writing on the blackboard, impromptu speech, instructional design and courseware, and simulated classroom instruction) while the other tests professional skills (basic knowledge about mathematics teaching and problem-solving).

Chinese scholars, researchers, and educators have provided frameworks and perspectives for studying and evaluating mathematics classroom instruction. For example, Gu and Zhou (1999) identified eight perspectives for studying mathematical teaching based on a review of Western and Eastern literature on classroom instruction: 1) content design, including teaching goals, the relationship between teaching requirements and students' learning, ways to construct knowledge and deal with key and difficult points, and the arrangement of instructional exercises; 2) instructional expression, including way to express concepts, language usage, and instructional strategies; 3) teaching schedule, including considering students' different characters; 4) materials used (e.g. technical equipment); 5) inspiring students' motivation through heuristic questions and exploring study through created learning situations; 6) interactions between teacher and students; 7) automatic learning; and, 8) encouraging students' creativity. Local researchers developed a scale (Wang, H., 2013) to evaluate Shanghai middle school mathematics teachers' professionalism (based on a related literature review, an investigation of the teachers, and their own teaching and research experiences), in which seven components were included: teaching objectives; teaching contents and the organisation of procedures; teaching methods; teacher basic skills; the relationship between teachers and students; teaching effectiveness; and students' learning.

The eight perspectives identified by Gu and Zhou (1999) encompass almost every consideration of mathematics classroom study. However, not every study should involve all eight, as different researchers will have their own research stances that may cause some of the perspectives to overlap. Wang's category of components emphasises teachers' developmental evaluation in teaching mathematics (Wang, H., 2013), but the detailed requirements of each component are hard to quantify as they rely on peer teachers' judgement. Furthermore, evaluation is generalised for mathematics teachers at different stages of expertise. The present study considers the above characteristics of mathematics teachers' teaching in China and adopts multiple perspectives to identify the main features of novice teachers' teaching practice.

2.3.3   Interactions among the Elements at the Individual Level

The previous sections reviewed individual teacher beliefs, knowledge, and practice in terms of their definitions, related studies and findings, and how to examine

them. It is believed that experience plays an important role in shaping the beliefs
and knowledge teachers bring into teaching. The complexity theory perspective
points to interactions among the aspects, rather than independent ones. According
to Opfer and Pedder (2011), dissonances may motivate teachers to learn or lead
to teacher change, such as dissonances between individual teachers' expectations
and sense of efficacy (Wheatley, 2002), or dissonances or conflict in teachers'
thinking (Ball, 1988; Cobb, Wood, & Yackel, 1990). According to Woolfolk Hoy,
Hoy and Davis (2009), such dissonance is 'change-provoking disequilibrium', and
only works when its extent is appropriate (Coburn, 2001; Timperley & Alton-Lee,
2008).

Some interactions among these aspects have been recognised in various studies. As mentioned in 2.3.1.1, Ernest (1989) pointed out that the relationship
between beliefs, knowledge, and teaching practice is that beliefs regulate the
other two. In particular, it is believed that the relationship between teacher beliefs
and their teaching behaviours is relatively complex (Schoenfeld, 2002). Based
on their review a substantial body of literature, particularly longitudinal studies (e.g., Kang, 2008; Mouza, 2009; Turner, Warzon, & Christensen, 2011),
Buehl and Beck (2014) identified four possible relationships between teacher
beliefs and practices: beliefs influence practice; practice influences beliefs; beliefs
are disconnected from practice; and reciprocal but complex relationships exist
between beliefs and practice. In this study, a reciprocal and complex relationship is regarded as a more proper interpretation. According to Buehl and Beck
(2014), relationship complexity may be strengthened by the variety in individuals'
features, contexts, and how beliefs and practices are assessed.

In addition, it is difficult to determine the roles that beliefs and knowledge
play in teaching (2.3.1.1). Teachers' professional beliefs may guide their practice, but their professional knowledge may constrain how they enact those beliefs
(Fennema & Franke, 1992). Ernest (2002) suggested that a reciprocal relationship exists between teacher beliefs and knowledge, with each contributing to
the other's renewal. Considering the reciprocal and complex relationship between
beliefs and practice, this study considers the multiple interactions among the three
aspects.

2.4   Influences of the School on Teacher Learning

It has been widely recognised that various aspects of the schools in which teachers are situated could support and constrain them, including the schools' norms,
structures, and practices (Galloway, Parkhurst, Boswell, Boswell, & Green, 1982;

Hollingsworth, 1999; Opfer & Pedder, 2011). From their comprehensive literature review, Opfer and Pedder (2011) summarised four organisational characteristics (focused on schools' processes and practices) that promote organisational and individual learning (p. 391):

O Nurturing a learning environment across all levels of the school (Barth, 1986; Hopkins, West, & Ainscow, 1996; Senge, 1990);
O Using self-evaluation as a means of promoting learning (MacBeath, 1999; MacBeath & Mortimore, 2001; MacGilchrist, Reed, & Myers, 2004; Rosenholtz, Bassler, & Hoover-Dempsey, 1986);
O Examining core and implicit values, assumptions, and beliefs underpinning institutional practices via introspection and reflection (Argyris, 1993; Argyris & Schön, 1978; Deal, 1984; Huberman & Miles, 1984; Senge, 1990); and,
O Creating knowledge management systems that leverage resources, core capabilities, and expertise of staff and pupils (Hargreaves, 1999; Leithwood, Leonard, & Sharratt, 1998; Marks, Louis, & Printy, 2000; Nickols, 2000; Nonaka & Takeuchi, 1995; Pedder, 2006; Rosenholtz et al., 1986; Zack, 2000).

Pedder (2006) suggested that schools' organisational learning optimises and sustains teachers' learning. Opfer and Pedder (2011) argued that the organisational learning mechanism is like an individual teacher system that involves various aspects and dissonance as motivators of teacher change.

School-level beliefs are considered to be the key influences on individual teachers' learning. These beliefs contribute to environmental norms that may impact both individual and collective behaviours (Sampson, Morenoff, & Earls; 1999). Based on the social theory of normative influences, Coleman (1985; 1987; 1994) claimed that individual teachers' practice is highly influenced by groups of teachers who share the same environment; novice or inexperienced teachers in particular more readily shape their practices to conform to collective beliefs (Chester & Beaudin, 1996; Woolfolk Hoy & Burke-Spero, 2005). After reviewing many studies, Tschannen-Moran, Salloum and Goddard (2014) confirmed that individual teachers are influenced by shared environmental norms that stem from the collective beliefs of the members situated in that environment and which work as key components of school culture. They also suggested two powerful norms that govern both individual and school practices: academic pressure (Hoy, Tarter, & Hoy, 2006; Berebitsky, Goddard, Neumerski, & Salloum, 2012;

Tschannen-Moran, Bankole, Mitchell, & Moore, 2013) and teacher professionalism (Tschannen-Moran, Parish, & DiPaola, 2006). As Tschannen-Moran et al. stated,

○ Academic pressure refers to the collective perception that there is a clear emphasis on academics in the school and that all students are to be held to high standards (Hoy, Hannum, & Tschannen-Moran, 1998);
○ Teacher professionalism, as an outcome and influencer of organisational beliefs, refers to teachers' perceptions that their colleagues take their work seriously, demonstrate a high level of commitment, and go beyond minimum expectations to meet students' needs.

For novice teachers, mentoring and induction provide a specific environment for the professional learning. This study will consider the influences of mentoring and induction on professional learning.

2.5 The Theoretical Framework of this Study

Based on the above literature review, this study adopts a theory that considers the complexity and dynamic of TPL and that is underpinned by both cognitive and situated perspectives on learning theory. For the cognitive perspective, three aspects (teacher beliefs, knowledge and teachers' practice) are investigated in terms of their definitions, frameworks or perspectives for assessing them, and how to measure them (2.3). In particular, the interactions among these aspects are hypothesised as motivators for teachers' learning. For the situated perspective, the study also takes into account the influences of the school (2.4), in particular environmental norms stemming from the collective beliefs of the members in the environment. It also examines the characteristics of novice teachers, particularly the stage theories of expertise, the features of Chinese novice mathematics teachers, and the influence of mentorship (2.2).

Finally, a conceptual framework (see Figure 2.1) was constructed to demonstrate the study's key concepts and their relationships. It demonstrates that 1) teacher beliefs, knowledge and practice (as defined in and being analysed through frameworks discussed in 2.3) are the main focal points of this study, as rich evidence of what novice teachers obtain through the teaching process; 2) the interactions among the three focuses are examined as the individual orientation to teacher learning (Opfer & Pedder, 2011); and 3) the influences of the school environment are also considered (2.4). It reflects from the framework that internal interactions among the individual elements and external interactions between individual teachers and their situated contexts construct a dynamic process of teacher learning. Consequently, learning outcomes achieved during the learning process indicate a new balance involving the relationships among the individual elements and between the individual teacher and the situated context.

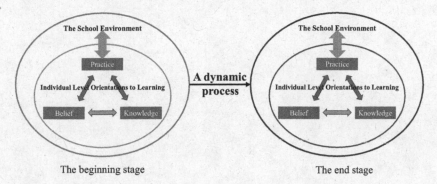

Figure 2.1   Conceptual framework of the study

Research Methodology and Study Design

As discussed in Chapter 2, the importance of taking into account the complexity and dynamics of teacher professional learning for education development has been widely recognised. Accordingly, this study focuses on individual teachers' perspectives and the influences of such different factors as teachers' experience, knowledge, beliefs, and practices, as well as the environment in which teaching is situated. With the explicit purpose of investigating the professional learning of novice mathematics teachers in the given context, this chapter justifies and describes the study's design. It places the study within a qualitative paradigm, indicates what sort of case study approach was adopted and why, describes the data collection process, illustrates how the data were analysed to address the research questions, and finally discusses the study's validity, reliability, and ethical issues.

3.1 Research Purpose and Research Questions

With an explicit focus on the professional learning of novice mathematics teachers at upper secondary schools in Shanghai, China, the study conducted a two-year investigation to produce a holistic image of the selected teachers' teaching in their situated location (including detailed descriptions of their beliefs, knowledge, and practice) and to explore how professional learning could be conducted. As mentioned in Chapter 1, two general questions were addressed in this study:

Supplementary Information The online version contains supplementary material available at (https://doi.org/10.1007/978-3-658-37236-1_3).

X. Lu, *Novice Mathematics Teachers' Professional Learning*, Perspektiven der Mathematikdidaktik, https://doi.org/10.1007/978-3-658-37236-1_3

1) What learning outcomes could the novice mathematics teachers obtain during the observed two years of teaching, particularly in terms of teacher beliefs, teacher knowledge, and teaching practice?
2) How was professional learning implemented to achieve the learning outcomes?

In answering these two questions, several sub-questions were also answered. The sub-questions related to the first question were as follows:

o What original beliefs and knowledge did the teachers hold when they began teaching?
o What features of their teaching practice in different time periods over the two years showed their learning?
o What aspects of their teacher beliefs, knowledge, and teaching practice did they change or keep consistent over the two years?

The second questions' sub-questions were:

o Based on the learning outcomes, what specific professional learning was performed by the teachers?
o What interactions among the teachers' previous experiences, beliefs, knowledge, and teaching practice could lead to or support their learning?
o What influences did the environment have on teachers' professional learning, especially as perceived by the teachers themselves?

To answer these questions, a qualitative research design was employed, as follows.

3.2   Adopting a Qualitative Research Design

Qualitative research is 'an umbrella term covering an array of interpretive techniques which seek to describe, decode, translate, and otherwise come to terms with the meaning, not the frequency, of certain more or less naturally occurring phenomena in the social world' (Van Maanen, 1979, cited in Merriam, 2009, p. 13). According to Yin (2010), qualitative research aims to 'study a real-world setting, discover how people cope and thrive in that setting—and capture the contextual richness' (p. 3–4). It is usually used to study phenomena in their natural settings and interpret them in terms of the meanings researchers bring to them

(Lincoln & Denzin, 1994). Compared with quantitative research, qualitative work allows researchers to better understand the multiple facets of phenomena within their specific contexts through the more focused description and analysis of a limited sample. A qualitative research design appears to be an appropriate means by which this study can develop a comprehensive understanding of the complex teacher professional learning process.

As Yin (2010, p. 8–9) concluded, qualitative research can 1) study the meanings of people's lives under real-world conditions; 2) represent the views and perspectives of the people in a study; 3) cover the contextual conditions within which people live; 4) offer insights into existing or emerging concepts that may help to explain human social behaviour; and, 5) use multiple sources of evidence rather than relying on a single source alone. Qualitative research can be used to investigate natural phenomena, can be conducted in naturalistic settings and without interventions (Denzin & Lincoln, 1994, p. 1), and should be both intense and prolonged (Dörnyei, 2007). Finally, the interpretation of research findings should yield thick, rich descriptions of sufficient detail to provide readers with a complete picture of the phenomena in question (Merriam, 2009). For example, qualitative research can be used to describe and report on the creation of key concepts and to generate and test theories (Cohen, Manion, & Morrison, 2007). The present study adopts a qualitative research design to obtain insights into the professional learning of novice mathematics teachers in the context of China from an insider's perspective without intervening in the investigation of the natural phenomena of teacher learning. The richer, more detailed data yielded by examining multiple aspects of teacher beliefs, knowledge, and practice as well as their interactions with their situated environment, would contribute to an in-depth understanding of teachers' teaching and learning. In this sense, the study may contribute to the complex and dynamic theory and practice of teacher professional development in the given context.

3.3 Case Study Research Model

3.3.1 Choosing a Case Study Approach

Instrument(s) selection frequently depends on an important earlier choice—which kind of research to undertake (Cohen et al., 2007, p. 83). The case study approach provides researchers with real examples of participants in real situations and enables people to understand ideas more clearly than abstract principles or theories (Cohen et al., 2007, p. 253). As defined by Feagin, Orum, and Sjoberg (1991,

p. 2), a case study is 'an in-depth multi-faceted investigation, using qualitative research methods'. In this study, a case study approach was selected to study the chosen object (Stake, 1994) by drawing on various techniques and procedures, including interviews, classroom observations, analysis of related documents, and field notes (Adelman, Jenkins & Kemmis, 1976).

The case study approach aims to contribute unique knowledge of various phenomena while allowing us to retain the holistic and meaningful characteristics of real-life events (Yin, 1994). The case study approach is a research strategy that focuses on 'how' and 'why' questions, requires few or no controls, and is based on a contemporary set of events. Hence, it has the following key features. First, it allows researchers to investigate in depth and to capture the holistic and meaningful characteristics of complex phenomena. Second, it gathers detailed information about the case (including its particular contexts) from multiple perspectives using various data sources. Third, it facilitates thick, objective descriptions of the case. Two main case study designs exist: single case and multiple case. Multiple-case design is more compelling and robust than single-case design but requires extensive resources and time (Yin, 1994).

The case study approach adopted in this study allowed intensive, holistic, empirical investigation of novice mathematics teachers' professional learning within their natural teaching context (i.e., the schools in which the teachers were situated). Additionally, the approach could contribute more to theory formation owing to its greater variation across cases. This study involved a case sample of three, which limited the multiplicity of the cases but allowed for in-depth analysis.

3.3.2   Participant Recruitment

To achieve the research purpose and answer the questions presented in 3.1, a longitudinal investigation was conducted to study selected cases of teachers' teaching in the two academic years from September 2013 to May 2015, with particular focus on such aspects as teacher beliefs, knowledge, and teaching practice. Four rounds of data were collected to track the professional learning process of the novice mathematics teachers using multiple instruments (detailed information in 3.4).

To address research questions with rich information, cases must be selected purposefully (Patton, 2002, p. 230). Given its particular focus on the beginning stage of teachers' professional learning, this study intentionally chose teachers who had just entered school to begin teaching. For richer information about contextual and individual factors, participants were expected to have various teaching and learning experiences, and to have worked in different schools. Eventually, the study recruited five novice mathematics teachers who had obtained bachelor's and master's degrees in mathematics or in education-related majors, had started their teaching at key-point upper secondary schools[1] in Shanghai in September 2013, and were willing to take part in the study. The procedure for recruitment of the teacher cases was as follows.

The selection of cases also used the researcher's convenience. In June 2015, the researcher contacted university educators, school principals, or experienced teachers in Shanghai, whom she knows directly or indirectly, with the aim of drawing on their connections to find potential participants. Meanwhile, she also learned from these educators that secondary and even primary schools in such a developed city in China—Shanghai—have precedence when recruiting highly educated candidates who graduated from prestigious universities. Thus, potential informants were sought from amongst those who had graduated in June with Master's or Bachelor's degrees, and had obtained employment in secondary schools in Shanghai. In addition, to control for the large number of factors that may influence teachers' professional learning, participants who would work at the same grade level were selected. Working in a key-point upper secondary school was another recruitment criterion, as most potential participants were bound for upper secondary schools; it was believed that key-point schools would be more supportive of teachers' professional development, owing to their high level of teaching quality; however, this constrains the study's generalisability. Five teachers were recruited at the beginning of the study. One quit during the second semester, having assigned by the school in which she works to study the teaching of IB curriculum abroad. Another case was not presented in this study owing to a unique feature of her school that would render it easily identifiable. Eventually, three potential informants with different teaching-related experiences took part— Doris, Jerry, and Tommy (pseudonyms). Their personal information is presented in Table 3.1 below.

---

[1] Key-point upper secondary schools in Shanghai are considered as top upper secondary schools in both student and teacher quality by local educational administration (descriptions see the website which retrieved on March 31, 2022: http://baike.baidu.com/view/14588853. htm).

Table 3.1 Basic
information of teacher
participants

	Gender	Educational background	School
Doris	Female	BSc (NU), MEd (NU)	School A
Jerry	Male	BSc (NU), MSc (CU)	School B
Tommy	Male	BSc (CU), MSc (CU)	School C

Notes: NU = normal university, CU = comprehensive university.

3.3.3 Research Context

To convey a better understanding of the informants and the teaching and learning
they delivered, some information about the research context is provided.

3.3.3.1 Educational Background of Novice Mathematics Teachers in China

Pre-service teacher education in China is provided by multiple institutions,
including three-year normal schools (which accept lower secondary graduates),
normal specialised postsecondary colleges (for upper secondary graduates), four-
year normal colleges and universities, and comprehensive universities offering
four-year B.Ed. programmes (Fan et al., 2014; Li, Zhao, Huang, & Ma, 2008).
As principals now have more decision-making power regarding staff, schools in
Shanghai prefer to employ graduates from universities rather than from closed-
system teacher training programmes (Fang & Paine, 2000). Moreover, schools
tend to judge teachers' ability based on their degree attainment, and rely on it
to improve teacher professional development (Gu, 2006; Wei & Bao, 2009); as a
result, a growing number of novice teachers now hold master's or even doctoral
degrees.

All the teachers who participated in this study have both bachelor's and master's degrees from normal or comprehensive universities in either mathematics
or education[2] (see Table 3.1). A Bachelor of Science (BSc) in the mathematics
programme at a normal university is different from one at a comprehensive university. Normal universities, as traditional teacher-preparation institutions, offer
courses relating to both mathematics and pedagogy (Fan et al., 2014, p. 49), and
their pre-service teachers attend teaching practicums; comprehensive universities,
on the other hand, focus on teaching mathematical content knowledge (particularly advanced mathematics), and their graduates do not, as Tommy reported,

---

[2] This is a basic mathematics teacher recruitment criterion for the secondary schools in which
the informants work.

receive actual teaching training. Nonetheless, comprehensive university graduates are popular teacher recruitment candidates in many schools in China, as they are believed to have more specialised knowledge of mathematics as a subject, and pedagogical knowledge may be obtained during teaching practice (Hu, 2001). Since 2012, the educational bureau in Shanghai has offered a unified induction programme aimed at providing normative teaching practice training for newly recruited novice teachers.

3.3.3.2 Induction Programme for Novice Mathematics Teachers in Shanghai

The induction programme, called 'normative training for first-year teachers at elementary and secondary schools (including kindergartens) in Shanghai'[3], is organised by the Shanghai Municipal Education Commission, which is responsible for local education policies and the direction of Shanghai's education system. The Commission drafted the guideline for normative training[4] in 2012, and the further guideline for normative training[5] in 2013. The guidelines pointed out that establishing normative training for first-year teachers was consistent with national and municipal standards for medium- and long-term reform and development in education (2010–2015)[6] and was intended to improve the teacher management system and promote teacher team construction in Shanghai's basic education system.

All participants in this study were involved in such an induction programme in the academic year 2013/2014. As the programmes were independently implemented by the education bureau at the district level, the cases' involvement was considerably similar but not wholly identical. The programmes provided abundant school-based (both intramural and interscholastic) teaching activities to allow new teachers to obtain information from or communicate with their more experienced peers. One important programme activity was mentoring; the programme designated certain key-point schools in each district as base schools, and assigned

---

[3] The original Chinese name is '上海市中小学 (幼儿园) 见习教师规范化培训'.

[4] The original Chinese name of the guideline is '上海市中小学 (幼儿园) 见习教师规范化培训的指导意见'; the document can be downloaded from the Shanghai Municipal Education Commission website (retrieved on March 31, 2022): http://edu.sh.gov.cn/xxgk2_zdgz_jsgz_02/20201015/v2-0015-gw_406112012007.html

[5] The original Chinese name of the further guidelines is '进一步推进上海市中小学 (幼儿园) 见习教师规范化培训', and the document can be downloaded from the Shanghai Municipal Education Commission website (retrieved on March 31, 2022): http://edu.sh.gov.cn/xxgk2_zdgz_jsgz_02/20201015/v2-0015-gw_406112013013.html

[6] 国家和上海市中长期教育改革和发展规划纲要 (2010–2015)

each novice teacher two mentors—from their own school or a district-based school—who guided the mentees' mathematics teaching and classroom management. Since the participants in this study worked in designated base schools, their mentors undertook mentoring tasks in both their schools and other schools. Although informal mentoring has always been implemented in Chinese schools, the three participants noted that the programme made the mentoring more normative or formalised by specifying how many times mentors should observe the mentees' classroom instruction and discuss their instruction with them. Doris thought that the rules for mentoring required by the programme were not useful, as the positive relationship between her and her mentor allowed her to communicate with him freely; Jerry perceived the mentoring as normative under the induction programme, since it guaranteed that they would receive mentor feedback on a set number of occasions.

Other activities, such as lectures by experienced teachers or educators, were considered useless in all three cases, as they were too dogmatic or generic to be practical. In addition, they complained that the programme handbook—in which they carefully documented activities they attended, their sample lesson designs, the journals they read, and other programme-related activities—was an unnecessary waste of time. Generally, the three participants thought that the induction programme was useless overall, except for the actual mentoring; however, as that occurred in their own schools and had always existed, they did not associate it with the induction program. As such, this study did not place particular emphasis on the induction programme but focused instead on specific environmental supports or constraints (e.g., mentoring) reported by the informants.

In addition to the government-required induction programme, schools may also provide teachers with opportunities to familiarise themselves with daily teaching practice: Jerry and Tommy were required to spend a semester learning about teaching before being allowed to start teaching in their new schools, as will be detailed in Chapters 5 and 6.

3.3.3.3 Teaching at Upper Secondary Schools in Shanghai

When conducting its investigations, a public upper secondary school in Shanghai usually includes different departments—for example, a regular department, an international department, and a Xinjiang department—that oversee student development. The regular department usually services Shanghai students who will sit the national college entrance examination; the international department is for foreign nationals learning an international curriculum (e.g., the IB curriculum); and the Xinjiang department enrols students from the province of Xinjiang to help promote education of China's national minorities. Owing to the departments' different purposes and their students' different characteristics, teaching in the departments is diverse in terms of teaching contents, teaching progress, and so on. While recruiting study participants and collecting data, it was found that novice teachers did not have the freedom to decide which category of students they would teach. To control for the number of factors influencing teachers' professional learning, the study focused on potential informants working in the schools' regular department, in which the Shanghai curriculum is implemented. However, one of the three cases, Tommy, was still assigned to teach a Xinjiang class in his second year of teaching, which made his second-year teaching different from that during this first year (details see Chapter 6).

3.4   Data Collection

This study adopted a case study approach to investigate the professional learning of novice mathematics teachers in Shanghai and employed multiple data collection methods, including semi-structured interviews, classroom observation, field notes, and document review. This section will detail the data collection procedure and instruments.

3.4.1   Data Collection Procedure

As mentioned in 3.3.2, data collection was conducted over four semesters in the academic years 2013–14 and 2014–15, with each round being conducted in the corresponding month of a corresponding semester; for example, the first round was in the first month of the first semester, and the second round in the second month of the second semester. The data collection sequence allowed a comprehensive investigation of the participants' development by sampling different time points in the two-year research period. Interviews, classroom observation, a review of relevant documents, and field notes were used in all four rounds to collect data on the teacher cases. Additionally, in the final round, interviews were conducted with the teachers' mentors (where possible) as a means of understanding their pedagogical beliefs and their views on their interactions with the participants, particularly with respect to mentoring. Detailed information is presented in Table 3.2, and the instruments used are discussed in the subsections that follow.

Table 3.2 Data collection schedule[7]

Time	Data collected	Aims
Sept, 2013	Interviews about general personal and school information Classroom observation of three to four consecutive lessons of each informant, documents related to teaching, interviews about the preparation, implementation, and reflection of the lessons Interviews about beliefs and interviews for 'discussing mathematical problems'	Knowing about personal and school background Investigating main features of the teacher cases' daily teaching, and beliefs and knowledge reflected in teaching at the beginning stage Investigating the stated beliefs held by the informants at the beginning stage, as well as the knowledge they had
March, 2014	Classroom observation of three to four consecutive lessons of each informant, documents related to teaching, interviews about the preparation, implementation, and reflection of the lessons	Investigating main features of the informants' daily teaching, and beliefs and knowledge reflected in teaching at the medium stage
November, 2014	Classroom observation of three to four consecutive lessons of each informant, documents related to teaching, interviews about the preparation, implementation and reflection of the lessons	
May, 2015	Classroom observation of three to four consecutive lessons of each informant, documents related to teaching, interviews about the preparation, implementation and reflection of the lessons Interviews about beliefs, interviews for 'discussing mathematical problems' Supplementary interviews for the mentors about their beliefs and views about the mentoring, and for mentees about their own reflection on their two-year's teaching	Investigating main features of the informants' daily teaching, beliefs, and knowledge reflected in teaching at the end stage Investigating informants' initial stated beliefs and knowledge Triangulating information provided by the novice teachers, especially their interactions with mentors

---

[7] Related documents and field notes were also collected during the process if necessary.

3.4.2 Classroom Observation

Observation allows researchers to gather 'live' data in naturally occurring social situations (Cohen et al., 2007, p. 396). Systematic classroom observation is a research method used to gather data on behaviours and interactions in classrooms (Croll, 1986). According to Croll, the method has various purposes (such as describing classroom features and monitoring individuals or teaching approaches) with respect to teacher development and initial training. Both quantitative (e.g., questioning techniques) and qualitative ethnographic traditions can be used in classroom observation. Ethnography is a process-oriented approach that explores what happens in the classroom by treating the classroom as a cultural entity (Van Lier, 1988).

The present study includes both quantitative and qualitative analyses of transcripts and employed an alternative perspective on ethnography (Holliday, 1997) that was process-driven but less in-depth than is typical for ethnographic research. Through nonparticipant observations, four classroom observation cycles (see Table 3.2) were implemented to investigate the main features of novice teachers' teaching practices, their beliefs, and the knowledge reflected in their teaching. The corpus of 38 observed lessons (list in Appendices 8.1.1, 9.1.1 and 10.1.1) were all video-recorded and analysed (see 3.5).

3.4.3 Interviews

An interview is a technique that allows participants to openly discuss their interpretations of the world in which they live and freely express their perspectives on specific situations (Cohen et al., 2007). Semi-structured interviews were used in this study to allow participants to express their opinions with help and guidance from the interviewer. The technique allows the researcher to elicit answers to a standardised schedule of questions but to do so in an informal manner in which the sequence of the questions may be freely modified and the questions themselves re-worded, explained, expanded, or added to, as needed (Cohen et al., 2007, p. 351). Interviews have certain disadvantages, including that the interviewees might offer responses that only reflect what they think the interviewer wants to hear; that are inarticulate, unperceptive, or unclear (Creswell, 2012). Thus, interview items must be carefully developed and piloted, and the researcher should be careful to say as little as possible during the interviews, to handle interviewees' emotional outbursts, and to use icebreaking techniques to encourage participants to talk freely (Creswell, 2012).

Interviews are typically used to access an informant's knowledge and beliefs, collect qualitative information relevant to the research questions, explore variables and relationships in the quantitative data, and facilitate data triangulation (Cohen et al., 2007). For the specific purposes of this study, five categories of semi-structured interviews were conducted during the research period (see Table 3.2), four of which directly involved novice teachers. The first (in the first round) collected novice teachers' personal information and basic information about the schools in which they worked (Appendix 1). The second category, which addressed their beliefs (Appendix 2) and a task of 'discussing mathematical problems' (Appendix 3), was carried out during the first and fourth rounds of data collection to identify their stated beliefs and knowledge at the beginning and end of the two-year research period. The third category, usually conducted immediately before or after classroom observations (Appendix 4), was used to enrich the understanding of teachers' classroom teaching practice, including how they prepared for and reflected on the lessons. The fourth category, conducted at the end of the two-year period, gathered teachers' reflections on their learning for teaching (Appendix 5).

The fifth and final interview category involved the novice teachers' mentors and addressed the mentors' pedagogical beliefs and reflections on their interactions with their mentees (Appendix 6) for the purpose of triangulating the data provided by the teacher cases. The contents and sample questions of all five interview categories are listed in Appendices 1 to 6. All interview data were transcribed verbatim and analysed in a qualitative way (described in 3.5) to achieve the above-stated purposes.

3.4.4 Document Collection

Documents are also considered a valuable data source in qualitative studies, since they can provide information that helps us understand key phenomena (Creswell, 2012). Numerous public and private documents were collected during the two-year investigation. Public documents (see Appendix 7), including policies and regulations about teacher education in Shanghai or the case schools, were collected to facilitate understanding of the research context; teachers' and students' books, curriculum documents, syllabi, and other teaching materials were also collected, including lesson plans, slides, assignment sheets, and reflections on class assignments. Such documents can yield useful data, as they usually reflect participants' thoughts in their own language and words and are ready for analysis without the need for transcription. Not all the documents collected are directly cited in the study; some were used to supplement the understanding of the informants' teaching behaviours and situated environments.

3.4.5 Field Notes

Field notes provide a holistic description of the research setting, including the people involved, their reactions and interactions with others, noteworthy events, and researchers' own reflections (Ary, Jacobs, & Razavieh, 2002). In this study, field notes were used to record information from the research setting, including observations about the schools and classrooms, teachers' casual conversations with their colleagues or the researcher, the researcher's reflections on the classroom observation, interactions between and among the cases and their colleagues, and her communications with the cases. These field notes contributed to the integrity of the data set as a whole and the researcher's preliminary analysis of the data.

3.5 Data Analysis

In a qualitative research paradigm, data are analysed inductively (Johnson & Christensen, 2000), and theory is expected to develop in a 'bottom up' way (Bogdan & Biklen, 1997). In this study, data analysis began with the data collection process and was guided by the broader research design, since hypotheses may be established and refined over time as data collection and preliminary data analysis proceed (Simons, 2009). Preliminary data analyses help to identify significant, critical, or even anomalous phenomena that occurred during data collection; the researcher's thoughts, reflections, comments, and suggestions regarding these phenomena are also recorded. Interesting and significant data perspectives were identified for further analyses involving the review of additional related literature.

Table 3.3 Summary of data analysis

Stage	Data analysed	Aspects of findings
Beginning Stage	Data collected in September, 2013	General background information of the informants and their schools Main features of the teaching practice, including lesson preparation, implementation and reflection Teacher beliefs, including stated beliefs and enacted beliefs at the beginning stage Teacher knowledge, including knowledge examined in 'discussing mathematical problems' and reflected in teaching at the beginning stage
Medium Stage	Data collected in March and November 2014	Main features of the teaching practice, including lesson preparation, implementation and reflection Enacted beliefs Teacher knowledge reflected in teaching
End Stage	Data collected in May 2015	Main features of the teaching practice, including lesson preparation, implementation and reflection Teacher beliefs, including stated beliefs and enacted beliefs at the end stage Teacher knowledge, including knowledge examined in 'discussing mathematical problems' and reflected in teaching at the end stage The informants' own reflections on the two-year's professional learning The mentors' stated beliefs and their reflections on the interactions with the informants (if applicable)

Data analysis was process-oriented and underwent several changes in focus. Initially, for example, mentorship was seen to play an important role in the early stages of teachers' learning; then, significant commonalities were observed when analysing the teacher beliefs, necessitating a change in the study's analytical focus. Consistent with the dynamic and complex nature of teacher professional learning, the study then focused on differences in the participants' individual professional learning processes over the two years, before discussing the three cases together to facilitate comprehensive understanding. Consequently, mentoring was found to be an important factor influencing teachers' professional learning, and common features were observed in learning outcomes and means of achieving learning, reflecting the impact of the teaching environment. With a particular focus on a holistic description of participants' professional learning processes, aspects of teachers' teaching practices, beliefs, and knowledge were analysed to assess their teaching and learning status at different stages of the two-year research period (see Table 3.3).

Generally, analysis of each aspect was theory driven, and various theories, frameworks, and analytic strategies were involved. Further analysis of the combined findings from the analyses of the individual aspects also followed related theories on teachers' professional learning. The selection and combination of the focal aspects for analysis was more data-driven and contributes to the complex and dynamic theory of teacher professional learning.

3.5.1 Analysis of the Three aspects in Each Stage

3.5.1.1 Teaching practice

As mentioned in Chapter 2, teaching practice is an essential, observable aspect of teacher education that is interrelated with other aspects. Researchers choose different perspectives for different purposes to explore teachers' teaching practices (e.g., LPS, Clarke, Emanuelsson, Jablonka, & Mok, 2006). With a specific focus on the professional learning of novice mathematics teachers in schools, this study took a holistic but less in-depth look at teachers' classroom teaching, involving the main features of basic teaching tasks completed pre-, in- and post-class (Reynolds, 1992).

Pre-class task: Preparing lessons
The study aims to understand how novice teachers prepared lessons at different stages during the two-year research period. Semi-structured interviews were conducted for the purpose of asking related questions (see Appendix 4) during each round of data collection (either before or after classroom teaching observation) so that informants could answer questions easily by associating them with recent teaching contexts. Analysis of the interview transcripts yielded rich data on various aspects, including how teachers prepared their lessons, environmental requirements or constraints (Lesson Preparation Groups), support from or interactions with other teachers, teaching materials used, and teachers' foci during the preparation process.

In-class task: Implementing lessons

Classroom observation data were primarily analysed to reveal the main features of lesson implementation. Various techniques (e.g., discourse analysis) have been used to analyse multiple aspects of classroom instruction; for example, the TIMSS video study investigated lessons' structures, mathematical content, and the mathematical problems used in classroom learning and teaching of mathematics (Hiebert, Gallimore, Garnier, Givvin, Hollingsworth, & Jacobs, 2003). Researchers have focused on particular features of mathematics teaching in China, such as teacher-centredness and student-centredness (Huang & Leung, 2005), as well as instructional coherence (Chen & Li, 2010). To demonstrate the main features of the novice mathematics teachers' classroom teaching, this study analysed several aspects identified in various related theories, frameworks, or analytic strategies. Combining the findings from the three cases, it eventually narrowed its primary focus to three aspects: the structure of the observed lessons, in-class questioning, and an overall impression of the teachers' teaching of the lessons.

The structure of the lessons. Mathematics lessons in China have generally been found to be well structured (Fan et al., 2014), and lesson structures can reveal such organisational features as instructional components and foci. For instance, in analysing mathematics lessons in Shanghai, Lopez-Real, Mok, Leung, and Marton (2004) observed that three activities (foundation/consolidation, exploration and guided practice) occupied the entire instruction, of which guided practice appeared to be the most significant, based on the amount of time allocated to it.

To demonstrate how the teachers implemented their classroom instruction, the study analysed lesson structure with a particular focus on content and process, as other researchers have in the past (e.g., Stigler & Hiebert, 1999). First, one representative lesson was selected for each case from each data collection round, based on when central topics were taught. Following Chen and Li's (2010) method for analysing instructional coherence, the selected lessons were then partitioned into activity segments. The relationships between the activity segments were coded to demonstrate coherence, the purpose of the segments (e.g., reviewing, introducing, and closing), the main form of discourse (e.g., teacher demonstration, teacher/student interaction), and the mathematical examples involved (see figures in Appendices 8.1, 9.1, and 10.1).

Questioning. Preliminary analysis of the questions posed in the observed lessons revealed that the teachers asked an enormous number of questions to facilitate highly interactive classroom instruction, as also found by Fan et al. (2014, p. 58). Additionally, it was found that more than 90% of questions were asked by the teachers and that the questions students posed mainly sought teachers' acknowledgement of their questions or answers to their questions. Thus, the study focused on teachers' questions, which followed Graesser and Person's (1994) definition. The questions posed in each lesson were coded based on their purpose (Moyer & Milewicz, 2002), the types of answers received (Graesser & Person, 1994), and who responded (see Table 3.4). During coding, teacher/student classroom dialogues were also analysed to see whose authority in initiating the questions and deciding the correctness of the answers. The analysis contributed to our knowledge of teachers' questioning strategies.

General features of the teachers' classroom teaching. Based on the above analyses of classroom instructions, the researcher summarised general features of the teachers' teaching when analysing and re-analysing the recorded videos, and the researcher's significant impressions of the observed lessons that were recorded in field notes also contributed to her interpretation.

After-class task: Reflecting on classroom teaching
Reflection is considered a key activity in teachers' learning (Postholm, 2012). Interviews were conducted to ask teachers how frequently they reflected on teaching and what they focused on while reflecting. These two points were interpreted in the study as demonstrating how the teachers perceived their daily teaching.

The above descriptions show how we investigate teacher practice. Regarding the interactions between teaching practice and teacher beliefs and knowledge, the teachers' classroom teaching was also analysed in terms of enacted beliefs and knowledge reflected in teaching, as described in the subsections that follow.

Table 3.4 Coding scheme for teacher questioning in class

	Code	Description
Answer to the question	• Yes/No	The answer to the questions is yes/no for verification.
	• Factual	The answer to the question is usually basic facts, such as disjunctive, concept completion, feature specification, and quantification.
	• Open-ended	The answer to the question is expected as long and uncertain to invite students' explanation or exploration.
	• Not math-related	The question is not related to mathematics.
Purpose of the question	• Checking	Where the teacher proceeds from one question to the next with little regard for students' response, which includes no follow-up questions and questions with verbal checkmarks.
	• Instructing	Instructing rather than assessing, which includes leading questions that direct the students' response and abandoning questing and teaching the concept.
	• Probing	Inviting or further investigating the students' answers, which includes questioning only the incorrect response, non-specific questioning and competent questioning.
	• NA	For non-math-related questions.
Responder to the question	• Individual students	A certain student answered the question.
	• Whole class	Several students or the whole class answered the question together.
	• No response	No students answered the question.

3.5.1.2 Teacher Beliefs

Researchers have suggested that a comprehensive understanding of teacher beliefs cannot rely solely on teachers' own reported beliefs but also requires examination of the enacted beliefs observed in teachers' instructional practice (e.g., Speer, 2005; Thompson, 1992).

Table 3.5   Teachers' stated beliefs

	Teachers' views	Examples
Beliefs about the nature of mathematics	The features of mathematics The components of mathematics The application of mathematics	The facts and results of mathematics are definite: right is right, and wrong is wrong (Doris-R1); mathematics is logical and systematic (Tommy-R1) Concepts, formulas, and the derivation of the concepts and formulas (Doris-R1); concepts, problem-solving strategies and mathematical thinking (Jerry-R1) Mathematics is the basic instrument of all scientific fields (Doris-R1); mathematics is a language used to translate real problems into mathematics and then solve them with mathematics (Jerry-R1)
Beliefs about the learning of mathematics	Students' ability to learn mathematics Effective ways to learn mathematics Important parts of learning and learning outcomes	Most of the teachers believed that students have different abilities or competences in learning mathematics Good learning habits (Doris-R1); Thinking when doing exercises (Doris-R4); Learning to summarise mathematical problems and correct errors made in exercises regularly (Jerry-R1) The teachers all considered students' understanding of mathematics as important, and emphasised more about the academic scores students had
Beliefs about the teaching of mathematics	Goals of mathematics teaching Effective teaching methods Attributes as a good mathematics teacher	These are usually consistent with the learning goals and learning outcomes Forcing students to do plenty of practices and keeping grasping at students' performance in doing exercises and tests (Jerry-R4) Being able to explain mathematics in a clear and simple way (Doris-R1)

Stated beliefs

Teachers' self-reported beliefs are called 'stated' or 'professed' beliefs (Speer, 2005). Semi-structured interviews were used to examine the beginning teachers' stated beliefs at the beginning and end of the two-year research period. Analysis of the novice teachers' stated beliefs was mainly theory-driven, and followed the framework summarised in Chapter 2 (Table 2.2). Teachers' various views were abstracted through preliminary analysis of the interview data based on three belief perspectives on the nature, the teaching, and the learning of mathematics (see Table 3.5) and then categorised as either traditional, non-traditional, or a mixture of traditional and non-traditional.

Enacted beliefs
As mentioned in Chapter 2, teacher beliefs can be reflected in such instructional aspects as textbook use and the teachers' role in classroom instruction (Ernest, 1989), and in whether teachers emphasise students' performance and speed, or their understanding and efforts (Stipek, Givvin, Salmon, & MacGyvers, 2001). The analysis of the teachers' enacted beliefs referred to the framework presented in Table 2.2, Chapter 2. Analysis began by reviewing the teaching practice data (including classroom observation data and related pre- and post-interviews) for each case in each data collection round, and then making notes to summarise the kinds of beliefs reflected in the teaching practice data. For example, when reviewing Tommy's statement (at the beginning) about how he prepared lessons, the note 'Emphasis on contents' was made, meaning that Tommy emphasised contents in preparing lessons. The notes provided an aide memoire that not only contributed to the interpretation of enacted beliefs but could also be cross-referenced to the primary data, including classroom observation and field notes.

3.5.1.3 Teacher knowledge
As mentioned in 2.3.1.4, Chapter 2, two frameworks—MKT and KQ—are used to analyse teacher knowledge in this study. MKT helps the researcher have a better understanding of what types of knowledge were held by the teachers or reflected in their teaching practice, while KQ provides helpful dimensions (e.g., transformation and connections) to analyse the teachers' knowledge in teaching. The two instruments used in examining teacher knowledge are interviews on 'discussing mathematical problems', observations of classroom instructions.

'Discussing mathematical problems'
'Discussing mathematical problems' (说题) is a typical task used in the recruitment and assessment of mathematics teachers in China, in which teachers are required to declaim their understandings of specific problems relating to the curriculum, students, mathematical contents, etc. The problems used in the task are usually not difficult but are commonly encountered in regular exercises or tests or when synthesising several mathematical topics. Based on these criteria, this study selected two questions from *yi-ke-yi-lian* (一课一练, *One Lesson, One Exercise*), a widely used practice book in Shanghai that has been translated and adapted internationally (Fan, Xiong, Zhao, & Niu, 2018); the translations of the questions

are presented in Appendix 3. Analyses of the interview revealed which aspects the teachers emphasised and what types of knowledge were reflected in their discussion of the problems—including mathematical knowledge, knowledge of content and students, knowledge of content and teaching, and knowledge of curriculum (see Table 2.3).

Knowledge reflected in teaching
The analysis of teacher knowledge reflected in teaching was also theory driven, and followed KQ, a theoretical framework used to analyse pre-service mathematics teachers' knowledge. The framework consists of four dimensions (foundation, transformation, connection, and contingency) covering most situations in classroom teaching (see Table 2.4). After preliminary analysis of the observed lessons, the analysis placed particular focus on knowledge-in-action, which includes the transformation and connection dimensions, since the novice teachers emphasised transmitting their mathematical knowledge to students in a coherent and smooth manner. Teacher beliefs—an important component of the foundation dimension—are discussed separately in this study. In the transformation and connections dimensions, nine codes were used to analyse teachers' knowledge of teaching (Rowland, 2013) and to code the transcripts of observed classroom instruction. Selected episodes were then used to illustrate the features of teachers' use of knowledge. Additionally, related post-class interviews were also analysed to learn more about why and how the knowledge was used.

3.5.2 Further Analysis

Based on data analysis of the three aspects, the study presents a holistic description of novice teachers' teaching in the two-year research period. To focus on teachers' professional learning, however, the relationships among these aspects must also be considered. As suggested by Woolfolk Hoy et al. (2009), the conflictual interplay between individual teachers' experiences, beliefs, knowledge, and practice within their orientation towards learning may provoke changes in their teaching practice. Furthermore, it is widely recognised that schools can both provide support for and impose constraints on teachers in terms of norms, structures, and practices (e.g., Galloway et al., 1982; Woods, Jeffrey, Troman, & Boyle, 1997). Thus, this study first analysed and discussed the consistencies and inconsistencies between teacher beliefs, knowledge, and practices, based on the holistic description of the teachers' teaching during the research period. Simultaneously, the schools' support for or constraints on the teachers were also summarised. Teachers' interactions within their individual orientations towards learning were internal, while their interactions with the school environment were external. Combining the analysis of internal and external interactions at each stage revealed the dynamic process underlying their professional learning. However, the external interactions must be interpreted with caution because they were analysed mainly from the individual teachers' own perceptions.

During the exploration, the researcher could restructure her interpretation of the data in consideration of the relationships between the various aspects of the data analysed to make it more coherent. Both primary and coded data were read repeatedly by the researcher to ensure an integrated and credible interpretation.

3.5.3   Analysis Focusing on the Influences of Mentoring

Particular emphasis was placed on the influences of mentoring on novice teachers' professional learning. Kuckartz's (2014) text analysis method was applied to analyse the novice teachers' perceptions of the role of mentor, including the three category-based methods—thematic text analysis, evaluative text analysis, and type-building text analysis.

After a careful review of the related interview data, four categories of novice teachers' perceptions of mentoring were identified:

- Situation—shows the general descriptions of mentoring from the novice teachers;
- Observing—refers to topics raised when the novice teachers observed their mentors' classroom teaching;
- Observed—refers to the novice teachers' reflection on the discussions with their mentors after observing their classroom teaching; and
- Discussion or general reflection—presents the novice teachers' feedback, reflections, and comments on the discussion with their mentors on topics not related to observations (such as discussions on the problems they had in preparing the lessons), and their reflections on mentoring.

Referring to the two-dimensional model by Hennissen et al. (2008), the content and topics were analysed in the study, revealing the novice teachers' descriptions of the mentoring conversations about their own and their mentors' teaching. Moreover, analysis of the interview data also recognised the novice teachers' reflections on the nature of mentoring, such as what kind of activities they expected, the role of mentors in mentoring activities and the influence of mentoring on the novice teachers' professional learning as well as their attitudes and emotions towards mentoring.

With the above analyses, the study evaluated the input of the novice teachers and their mentors (i.e., whether the mentor or mentee were active in the mentoring activities), as well as the novice teachers' interpretation of the nature of mentoring and their attitude towards mentoring (i.e., whether the mentoring was supportive).

3.5.4   Analysis of Supplementary Data

To triangulate and enrich the researcher's interpretation of the novice teachers' professional learning, the study also collected supplementary data, consisting of interviews with the informants about their own reflections on learning to teach, and with their mentors about the mentors' pedagogical beliefs and their issues with the mentees (3.4). All interview data were transcribed and analysed, focusing on 1) what learning outcomes teachers thought they had attained during the two years and 2) what factors contributed to those learning outcomes, corresponding to the two research questions. The mentors' pedagogical beliefs were analysed in the same way that the teacher participants' stated beliefs were, as illustrated in 3.5.1.2. Regarding the mentorship issues reported by the mentors, the study mainly focused on 1) what instruction the mentor thought they had provided the mentee and 2) the mentors' comments on their mentees' learning, such as in what aspects they had improved or needed to improve.

3.6   Validity and Reliability

To have any effect on either practice or theory in a field, research studies must be rigorously conducted (Merriam, 2009). Validity and reliability are two common criteria for evaluating qualitative research studies. Internal validity (or credibility) concerns the extent to which research findings reflect reality; external validity (or transferability) concerns the applicability of the research findings to other contexts; and reliability (or consistency) concerns the degree to which the research findings can be replicated. To enhance the internal validity and reliability of the study, several strategies were used.

First, both data triangulation and methodological triangulation can strengthen internal validity and reliability (Denzin, 1989). This study used multiple methods of data collection (interview, classroom observation, document collection, and field notes) to investigate different aspects of and perspectives on teachers' teaching and learning for teaching. In addition, it also employed data triangulation. Four rounds of data collection—including a series of lesson observations and interviews—ensured that the phenomena under investigation were repeatedly observed; moreover, supplementary data, including informants' reflections on their learning of teaching over the research period and their mentors' comments, provided insider and outsider perspectives to triangulate the researcher's interpretations. For example, investigation of the mentors' beliefs as mathematics teachers helps confirm the kinds of collective beliefs that contribute to environmental norms influencing the novice teachers' teaching, and the mentors' views about their mentees' teaching also triangulated the participants' own perceptions of their learning.

Second, member checking was also employed to ensure validity (Lincoln & Guba, 1985). The interview transcripts, the researcher's main interpretations, and the tentative findings were reported to the informants and they agreed with the researcher's interpretations. Where any lack of clarity or misunderstanding emerged, the researcher provided further explanations and negotiated with the participants, and their opinions were solicited until agreement was reached.

Third, this study employed a complex and dynamic theory of teacher professional learning concerning various aspects of teachers' teaching as well as environmental influences and collected the richest data possible (Miles & Huberman, 1994). It also provided a clear explanation of the study's ethical considerations (3.7), the selection of informants (3.3.2), and the social context (3.3.3) in which the data were collected (Goetz & LeCompte, 1984).

Fourth, to facilitate other researchers' replications of the study (Geotz & LeCompte, 1984), a detailed audit trail was presented, covering such areas as data collection (3.4), data coding, and decisions made (3.5).

It is not expected that case study research can be extrapolated to a wider population. The major purpose of a case study is to understand a particular case by developing a thorough understanding of its uniqueness and complexity (Stake, 1988). However, as Eisner (1991) claimed, case studies are significant in that they carry implications beyond the limited contexts in which the cases are situated. Thus, this study also took generalisability (external validity) into consideration through strategies suggested by Merriam (1988), including the provision of rich descriptions to allow others sufficient information to judge its suitability for reference. In addition, the typicality of the cases was also discussed so that readers could have relevant references for their own situations. Although only three case teachers were examined, they were selected from different schools located in different districts of the research site, which enriches the findings' representativeness to a degree.

3.7 Ethical Considerations

With reference to Johnson and Christensen (2000), participants' informed consent and freedom in the research process, confidentiality, and potential risks and benefits are the major ethical considerations in studies of this nature.

For each participant involved in the study, informed consent was obtained prior to data collection. As the participants were mature adults, they were competent to take responsibility for their decisions. Before the study was conducted, potential participants were asked whether they were prepared and willing to participate in the study. It was made clear to the participants, when the research purposes and procedures were introduced to them, that their participation would not affect the evaluation of their teaching performance. Further details, such as the length and number of interviews and classroom observations and the types of documents to be collected, were explicitly explained to the participants. The participants also understood that participation was voluntary and was without any physical risk or emotional risk; they had the right to withdraw their consent or discontinue participation at any time and for any reason, without fear of any reprisals or consequences. They were assured that their personal information would be kept confidential throughout the study and in subsequent reports, and that the researcher would not intervene in the teachers' practices. If participants had any questions or concerns, they were invited to contact the researcher using the contact information provided to them on the informed consent form.

Regarding confidentiality, participants were ensured that all identifiers—including names, addresses, and affiliations—would be deleted or substituted with pseudonyms in any and all public research reports, and that participants' related background information would be reported generally, rather than specifically. Video-taping was used in the proposed study, and participants had the right to review the records and to erase any or all parts of the recordings if they wished. Video-taped classroom observation focused only on the teachers; however, since teachers usually walked around, students were sometimes captured on camera, in which case, a mosaic effect was used in the videos to mask their faces. Participants also had the right to decide whether to share the requested documents.

As this is a naturalistic case study, no risk of harm to the participants was likely. On the contrary, the participating novice teachers might benefit from having more opportunities to reflect on their teaching practice and from obtaining more information from their mentors during the study period.

The Case of Doris

4

4.1 Introduction

This chapter examines the professional learning of Doris (pseudonym), who worked in the key-point upper secondary school—School A, in District I—over the two academic years from September 2013 to June 2015. The investigation was conducted in September 2013 at the beginning of the two-year period, in March and November 2014—representing the medium stage—and during the final stage in May 2015.

Doris obtained a bachelor's degree in mathematics and applied mathematics and a master's degree in mathematics education (majoring in mathematics history and pedagogy) from a prestigious normal university in China. She had one year's teaching experience as a voluntary mathematics teacher as part of a programme promoting the development of education in China's western rural region. Investigation of her initial two years' teaching in School A revealed Doris's attempts to promote students' interest in learning mathematics that were limited by the necessity of adhering to the mathematics curriculum in the school, which prioritised students' examination results. Doris's learning focused on both teaching school mathematics and promoting students' interests and mathematical thinking in mathematics learning through the incorporation of mathematical history and culture into her teaching. The learning process was enriched not only by the interactions of her knowledge, beliefs and teaching practices but also by the constraints and opportunities provided by the school.

Supplementary Information The online version contains supplementary material available at (https://doi.org/10.1007/978-3-658-37236-1_4).

4.2   Background

4.2.1   Doris' Background

Doris considered her experiences during the four-year bachelor's teacher training programme to be useless because the opportunities for teaching practice in real school situations (i.e., practicum and internship[1]) were limited and took place long before she had the opportunity to teach in School A. She regarded her subsequent year of voluntary teaching experience in the west of China[2] as important for her teaching development. She explained that, in particular, it helped her to recognise what is important in the independent implementation of teaching in schools, such as each lesson's appropriate capacity[3] and pace, which requires the integration of various considerations, such as students' personalities, lesson contents and the practices of other teachers in the school. She also reported having experienced 'special feelings' about teaching in a different world that is rural and less-developed and that the culture and students there differed from those in Shanghai. Doris appeared to have been deeply moved by those feelings.

Doris also appreciated the three-year master's programme that she completed because it allowed her learn about the history and pedagogy of mathematics (HPM) and to equip herself with the consciousness and knowledge required to integrate mathematical history and culture into teaching. She remarked,

> I know what (mathematical) topics can be integrated with history, hmm, such as trigonometry. I also know what kind of materials [readings] to search for to prepare for integrating historical facts.

When Doris was asked about her reasons for becoming a teacher, she mentioned her particular preference for teaching, the influence of her own teachers during her childhood and her talent as an educator.

---

[1] The practicum usually lasts several days during the third year of study and allows student teachers to visit schools for the purpose of observing schoolteachers' classroom practices and asking questions or discussing teaching with the teachers. The internship lasts a semester for Year 4 student teachers, who observe schoolteachers' classroom teaching and deliver several lessons to students under the teachers' instructions.

[2] The program represents a response to calls for the construction of a new socialist countryside to promote the development of education in the central and western regions. The selected participants are outstanding normal university students who have a higher political consciousness and are likely to be admitted to master's programmes.

[3] 'Capacity' refers to the term '课堂容量' in Chinese, which indicates how many topics (or how much knowledge) should be taught in a single lesson.

From when I was very young, I, myself, I was that kind of student who liked learning. And I encountered several good teachers who impacted my love for teaching… I also thought that I had, hmm, some talent (for teaching). Anyhow, I just wanted to be a teacher from childhood and never thought about other careers.

4.2.2   School Information

The school Doris works in, School A, is a key-point upper secondary school that was established twenty years ago. The school comprises two independent departments—the regular high school department and the international department. Doris worked in the regular department, which consisted of three grades (Grades 10, 11 and 12), each of which had 12 classes in two categories—parallel classes and special classes.[4] The students in special classes typically exhibit better academic abilities than those in parallel classes and are enrolled based on their good performances in both the *Zhongkao* (the city's unified entrance examination for upper secondary schools) and the school's own entrance examination. With the exception of two extra lessons each week for the special curriculum, students in special classes are taught the same curriculum (the Shanghai Curriculum) as students in parallel classes. Doris taught two classes during the initial two years from Grade 10 to Grade 11—one parallel class (Doris-Class1) and one special class (Doris-Class2). The following analyses of Doris's teaching mainly use data from Doris-Class1.

Doris was involved in two lesson preparation groups (Doris-LPGs) that were relatively similar to one another across both years, as only one or two teacher members changed. The LPGs intervened in the group members' teaching by unifying students' homework and arranging or coordinating the schedule and pace of teaching. The unified students' homework was mainly selected from the students' exercise books, which had been edited by former LPGs from School A. The schedule for the entire semester was announced by the group leader at the beginning of each semester and was revised by the leader based on a schedule used by the LPG that taught the same content in the previous year. The schedule is a general plan that includes which mathematical topics will be taught and who is in charge of the weekly plans and progress tests.

However, Doris said the schedule was merely a formality, because 'the fact was that each teacher prepared and implemented their teaching by themselves'

---

[4] The special classes were established in accordance with a specific curriculum developed by School A to promote students' special literacy, such as foreign language, arts, science and mathematics.

and 'the most important function of it (the schedule) was letting school leaders know what we would teach in that semester'. She also noted that the occasional informal discussions[5] among the members provided teachers with opportunities to learn from one another and maintained a unified teaching pace across the whole group. According to Doris, the LPGs promoted or implemented collective teaching activities that were usually assigned by senior TRGs (teaching and research groups), such as the school TRG or the district TRG. Regarding the activities' organisation, the LPGs had some flexibility, and schools usually adopted their own approach to organisation. These activities could include, for example, particular rehearsals or discussions among group members with the aim of developing public lessons, unified test marking and reviewing so that teachers could compare their students' performance with others' as well as implementing specific teaching activities (e.g., 'same lesson; different structures'[6] in School A) so that the teachers could learn from one another.

4.3 The Initial Stage of the Two-year Professional Learning

As Chapter 2 described, this study employs a systematic conceptual framework to analyse teachers' professional learning and focuses on a nested system that includes both the internal and external interactions of the teacher and school subsystems (Opfer & Pedder, 2011). In this section, Doris's teaching process is first examined by exploring each individual element (e.g., teacher beliefs, knowledge and teaching practice) during each stage before discussing the interactions and relations between the elements as well as the interactions between Doris and the situation in which she was located.

4.3.1 Main Features of Doris's Teaching Practice

The main features of Doris's teaching practice are reflected in the activities she prepared, implemented and reflected on in relation to her classroom teaching. She designed the lessons independently with reference to several teaching materials and then made modifications in consideration of her students' features. She

---

[5] All LPG members shared the same office, and so they could have discussions at any time.
[6] 同课异构, which means that teachers teach the same contents in a lesson but use different designs.

implemented the plans with structured and teacher-controlled instructions and used school-developed materials to provide exercises for students. After class, she reflected on her day's teaching and, where appropriate, made improvements based on those reflections. Doris was generally satisfied with her teaching in view of her students' acceptable performance on the progress test.

4.3.1.1 Lesson Preparation

To ensure that she was well prepared to commence her teaching, Doris had written the plans for several lessons during the summer break. She stated that she could obtain a better understanding of the contents to teach in each lesson and make decisions on what and how to present the material in class, through studying the textbook and other supplementary teaching materials all by herself. As Doris said, in consideration of students' different academic abilities, she made differences in the lesson plans for the two classes (e.g., adopting different mathematical examples). However, as she began implementing her prepared lessons, she found that they were inappropriate for her students because she had underestimated their academic abilities. She thus modified or rewrote the plans, referring to the school-developed students' exercise book. She claimed that the exercise book contributed to her knowledge of the appropriate teaching pace and contents as it had been developed by school LPGs and integrated experienced teachers' rich knowledge about the students and the appropriate content.

4.3.1.2 Lesson Implementation

Doris implemented the classroom instructions in accordance with a clear structure that consisted of three parts: reviewing, introducing and closing (Appendix 8.1.1). She relied on the connections between the contents to construct the teaching procedure (ibid), and posed several easily answered questions (44.53% yes/no and 37.31% factual questions) to the whole class (50.85%) to engage them (33.27% for checking and 43.41% for instructing; for details, see Appendix 8.1.2). As Doris held the authority to initiate questions and determine whether the students' answers were correct, the classroom instructions appeared to be teacher-controlled (Li & Ni, 2009). To consolidate the students' knowledge acquisition, she also devoted considerable time to correcting mistakes in the students' homework and regular tests, which she believed helped her to learn more about the students within a short period.

4.3.1.3 Lesson Reflection

Doris reflected on her teaching after almost every class. She said that she usually 'had a general impression of it (the teaching): good or not'. These impressions represented a kind of 'inter-sensory judgement' based on her own feelings about the teaching process—whether it had been smooth and what messages she had received from the students in class, such as facial expressions indicating confusion or understanding. She recorded the problems (e.g., X is an inappropriate way to introduce Y content) she found in her reflection, considered ways to solve the problems, and trialled the solutions in the next class. She also reported that she was satisfied with her teaching at that moment and would continue it, since the teaching outcome—students' performances in a progress test—was acceptable.

> (Students' performances in the examination were) not bad. My current expectation (for the students) is that they will 'stand firmly on their feet (站稳脚跟)'.[7] They have achieved this already. I think this is fine—not the worst, nor the best.

4.3.2   Teacher Beliefs

During the initial stage, Doris's beliefs as a mathematics teacher were examined based on what she stated in the interviews (stated beliefs) and how she enacted in teaching (enacted beliefs).

4.3.2.1 Stated Beliefs

Researchers typically classify teachers' statements as their stated or professed beliefs (e.g., Speer, 2005) and analyse mathematics teachers' beliefs by examining their perceptions of the nature of mathematics and mathematics teaching and learning (e.g., Beswick, 2005).

The interviews (Appendix 8.2) indicated that Doris held mixed traditional and non-traditional beliefs, which is consistent with the view that mathematics is 'a static but unified body of certain knowledge; and expect for rules and procedures, there are concepts and principles "behind the rules"' (Ernest, 1989; Raymond, 1997; Thompson, 1991). Doris believed that mathematics consists of definite results or rules derived via various methods and that students should learn these results or rules as well as the ways in which they were derived.

Doris emphasised teachers' and students' efforts in mathematics teaching and learning. She considered the teacher's role to be that of an explainer and believed that teachers should have a solid subject knowledge base and sufficient pedagogical knowledge to clearly explain complex mathematical problems in easy ways. She also reported that rather than emphasising students' performances, she

---

[7] It means that students could consistently perform above the average level in the examinations among their schoolmates.

preferred to value their positive attitudes and understanding in learning mathematics. She considered HPM an effective means of promoting students' interests and active engagement in teaching and learning, a belief that she had obtained during her master's programme at university.

4.3.2.2 Enacted Beliefs

It is not sufficient for an investigation to rely solely on teachers' own reports (Speer, 2005); teachers' instructional practices or mathematical behaviours should also be examined to better understand their mathematical beliefs (Thompson, 1992).

Observation of Doris's classroom teaching and the follow-up interviews indicated that Doris's enacted beliefs were close to the mixed category. She paid considerable attention to students' understanding by focusing on lesson content, particularly since the lesson's structure was established based on the connections between the mathematical facts being taught.

However, Doris's teaching practice seemed more traditional. First, it appeared that she preferred to deliver teaching that adhered to the curriculum. The teaching materials she used to help students understand the contents and to make decisions about her teaching were highly related to the centralised curriculum—what Zhang (2012) would call supplements for well-designed textbooks. Second, in the classroom, she relied on teacher-controlled instructions to facilitate students' learning. Doris asked numerous easily answered questions to lead the teaching process, which included various mathematical examples to help the students grasp mathematical concepts and procedures. To promote the students' own efforts to learn mathematics, she asked them to preview new content before class as homework, checked their understanding of the contents and subsequently delivered her teaching based on their responses. However, her promotion of students' efforts was not significant as she focused only on students' attempts to understand a few simple concepts. She also emphasised students' performances in daily homework and on unified tests and assessed her own teaching based on the students' achievements in progress tests. As she reported in the post-class interviews, the first thing she had to do in teaching was to help students 'stand firmly' on their feet' as they competed with their schoolmates.

4.3.3   Doris's Teacher Knowledge

Doris' teacher knowledge and her use of this knowledge in teaching the task 'discussing mathematical problems' were analysed.

4.3.3.1 'Discussing Mathematical Problems'

The task 'discussing mathematical problems' is usually used in teaching and research activities in China to assess mathematics teachers' mathematical and pedagogical literacy (Huang, 2016). When the researcher delivered the task to Doris at the beginning of the two-year investigation, she cared greatly about the correctness of her solutions to the problems and considered the task to be a problem-solving procedure. She first summarised the main mathematical topics involved in the two problems from the perspective of examination—what she called the problems' 'exam foci'. She then discussed ways of solving the problems, emphasising the mistakes that students may make (see Appendix 8.3). Doris explained that she was used to considering 'exam foci' or 'error-prone points' in relation to such mathematical problems, as she had completed several similar exercises involving the kinds of questions asked in previous matriculation tests (*Gaokao*) while preparing to apply for teaching positions in Shanghai upper secondary schools; as she reported, teachers skilled in solving *Gaokao*-related mathematical problems are particularly sought after by schools.

4.3.3.2 Knowledge in Teaching

Doris's teaching knowledge was analysed from the two dimensions of *transformation* and *connection* (Turner & Rowland, 2011). Rather than haranguing, Doris preferred to use mathematical examples to introduce or demonstrate new content and to check or consolidate students' understanding of the content. For instance, to introduce the concept of 'imply', she used the mathematical example 'a natural number with a final digit of 5 is divisible by 5' to demonstrate the procedure of dividing a proposition into two parts: 'the last digit of a natural number is 5' and 'the natural number is divisible by 5'. At the same time, she invited students to participate in the construction of the concept 'imply' by asking them to justify whether the proposition was true (Appendix 8.4.1). Doris also employed a set of mathematical examples to introduce strategies for solving questions of the same type, such as finding the negative form of various propositions. The connections between the new content and previously learned knowledge became evident when she used these examples; for example, the mathematical topic 'inequality' was used to determine the negative form of the proposition 'there exists at least one positive number among a, b and c' (Appendix 8.4.2).

In the post-class interview, Doris reported that her teaching knowledge represented a kind of 'subconscious' instinct that she had obtained from her teaching practice, teacher training, and other personal experiences with schooling. She explained that it was difficult to say precisely what she might learn from her experiences with respect to specific teaching tasks; for example, she could not tell whether her understanding of which contents of a given lesson were particularly important or difficult stemmed from the teachers she had encountered as a student or the relevant exercises that she had completed. Moreover, for a more comprehensive understanding of the contents that incorporated her knowledge of the students, the curriculum and pedagogy, she still had to learn various teaching materials.

As Doris reported, the materials played different roles in her teaching knowledge. For instance, some supplementary teaching materials helped her to 'refine' her understanding of the content and draw connections with other mathematical concepts. Other materials and the textbook enriched her accumulation of mathematical examples to use in class, and the overlap of the examples used in these teaching materials informed her selection of mathematical examples. In particular, school-developed materials contributed more to her understanding of students' need and the knowledge of the teaching content. Doris implemented such knowledge in her teaching and paid considerable attention to students' acquisition of mathematical knowledge. For instance, when she found that students had difficulty understanding her previous teaching (e.g. 'sets'), she accelerated her teaching of 'proposition' to leave more class time to consolidate students' understanding of 'sets' by completing more exercises while maintaining the same teaching pace as other teachers.

4.3.4   Summary and Discussion

At the beginning of the two years, Doris prepared and modified lesson plans independently, using various teaching materials as references. With her teacher-controlled classroom instructions, the teaching she delivered appeared to adhere strictly to the curriculum materials. Reflecting on her teaching, she emphasised the smooth teaching process that was based on her own instincts and feedback from students. She was generally satisfied with her teaching because her students' performances on the unified progress test met her expectations. This demonstrated that she considered students' academic performance an important criterion in evaluating her teaching. She also emphasised both the content and the students' understanding of it. Various aspects of knowledge were reflected in her teaching, including knowledge of content, curriculum, students and teaching.

In her stated beliefs, Doris emphasised the content and students' understanding of the teaching and learning of mathematics, which was consistent with her mixed beliefs about the nature of mathematics. She particularly emphasised students' interests, attitudes and habits in the learning of mathematics. These seemed inconsistent with her teaching, which focused more on students' academic performance. In discussing mathematical problems, she appeared to focus particularly on the *Gaokao* examination.

4.3.4.1 The Individual-level Orientation to Learning

Following on from the above description of Doris' professional learning for teaching in the early stage, the section that follows discusses the interactions among individual aspects, including experience, beliefs, knowledge and teaching practice.

The influences of experience

Doris obtained her teaching-related experiences mainly through her personal experiences of school and instruction and from various teacher education programmes that she attended at a normal university. According to Richardson (1996, 2003), such experiences affect novice teachers' pedagogical beliefs and knowledge and thereby influence their teaching practice. Doris's four-year bachelor's programme seemed to contribute to her beliefs and knowledge about how well-designed curriculum materials should be delivered, which is consistent with Zhang and Wong's (2014) point of view; in addition to providing her with the relevant knowledge, Doris's master's programme inspired her to promote students' interest in mathematics by integrating mathematical history and culture into her teaching. The interactions between her experiences and her pedagogical knowledge may form what she called her teaching 'subconscious' and seems similar to what Gu, Wang and Yang (2014) called 'tacit knowledge'. Tacit knowledge denotes teachers' learning with respect to how to do and think, is particularly situated in the individual teacher's experience and is likely acquired through practice. Doris's comments on the effectiveness of the teacher training programmes also indicate that she emphasised the effects that rigorous, focused and timely training could have on her teaching, particularly independent teaching practice in real school situations.

Teacher knowledge and teaching practice

The integrated use of subject knowledge and pedagogical content knowledge was evident in Doris's classroom teaching; for instance, when she used 'inequality' (previous knowledge) to explain how to determine a proposition's negative form (Appendix 8.4.2), it revealed her knowledge about the content (e.g., connections with other mathematical topics), the students and the content (i.e., the difficulties students may have), and the curriculum and the content (that is, which content should be emphasised in the curriculum, particularly with respect to assessment).

The interviews revealed that Doris focused on her professional knowledge when teaching in School A. Based on the knowledge she had already obtained from her experiences, she seemed to understand which types of knowledge she needed, such as a comprehensive understanding of the contents and better understanding of students' needs. Her teaching practice directed her learning; when she noticed that the teaching she implemented was inappropriate for the students, she strove to learn more about the students by referring to the school-developed exercise book. She also paid attention to feedback from the students (including their responses in class and their performances in daily exercises and progress tests) for the purpose of assessing and adjusting her teaching.

This implies the reciprocal influences of teacher knowledge and teaching practice on Doris in the early stages of her teaching practice: her existing knowledge contributed to her teaching practice, and her daily teaching supported her in accumulating new, related knowledge.

Teacher beliefs, knowledge and teaching practice
Researchers have claimed that teachers' beliefs are the foundation of knowledge in teaching and include rules or grand ideas that guide other dimensions of knowledge in teaching, such as transformation, connection and contingency (Rowland, Turner, & Thwaites, 2014; Weston, Kleve, & Rowland, 2012). Doris, who believed mathematics to be static and the teacher's role to be that of an explainer, consistently implemented content-focused and teacher-centred teaching. Correspondingly, her use and accumulation of knowledge were also closely related to such beliefs and teaching practices. The above discussions on the interactions among teacher beliefs, knowledge and teaching practice suggest that teacher beliefs are the primary commonality between teaching knowledge and practice (Ernest, 1989).

On the other hand, Doris's specific belief in promoting students' interest by integrating history and culture into the teaching of mathematics was not enacted in her teaching. According to Woolfolk Hoy et al. (2009), dissonance between beliefs, knowledge and practice may result in 'change-provoking disequilibrium'. As such, could Doris be expected to learn anything new during the later stage of her two years of professional learning?

4.3.4.2 The School/organisational Level Orientation Towards Learning

Researchers have recognised the important role that schools play in teachers' professional learning. More than 25 years ago, Ernest (1989) highlighted the effects that the social context of the teaching situation had on teachers' teaching of mathematics, particularly the constraints and opportunities associated with the context. More recently, Opfer and Pedder (2011) claimed that the school, as a system, contributes to teachers' learning through its interactions with individual teachers. In her earliest teaching experiences, Doris immersed herself in relatively independent teaching practice and independently prepared, implemented and reflected on her teaching. Nonetheless, the school or social context continued to impact her teaching through the supplementary teaching materials and the unified teaching schedule, homework and progress tests. Through these, Doris could absorb the various elements of knowledge that she used in teaching each lesson, including her knowledge of content, the curriculum and the students. On the other hand, the contexts may also have constrained Doris' teaching by obliging her to focus more on students' performance than on their interest in mathematics, as when she struggled to help students to 'stand firmly on their feet' in the progress test so that they could compete with other students in the same grade.

4.4   The Medium Stage of the Two-year Professional Learning

An investigation of Doris's professional learning was conducted in March and November, 2014—the middle stage of the two-year period—to examine her teaching activities in the second and third semesters of teaching.

4.4.1 Main Features of Doris's Teaching Practice

During the second semester, Doris prepared lesson plans by referring to the collective lesson plans, which were written formally, and her own understanding of the content obtained during her specific learning experience. In the third semester, she learned from other mathematics teachers in her LPG through collegial discussions and by observing their classroom instruction. Although she still insisted on teacher-controlled instructions, she had more interactions with students after class to promote their exploration of mathematical problems. Reflecting on her lessons, she focused on her mentor's suggestions for good teaching behaviours and emphasised her own 'instinct' for teaching.

4.4.1.1 Lesson Preparation

During the second semester (March 2014), Doris still prepared lessons by herself but referred to lesson plans downloaded from the official Shanghai Educational Bureau website, as they 'were developed by lesson preparation groups of other upper secondary schools in Shanghai, and were written 'in a formal way', including structured lesson procedure, and key points and difficult points for each lesson'. She also appreciated her master's programme's contributions (in particular the literature on HPM) to her understanding of the specific teaching contents (*trigonometry*).

During the third semester (November 2014), Doris reported studying extra reading materials to understand how to integrate mathematical history into her teaching of the contents (e.g., *analytic geometry*). She particularly valued the opportunities to learn from other teachers in the same LPG by observing their classroom teaching—particularly the mathematical examples they used and how they introduced the contents. She expressed particular gratitude to her mentor, Teacher Zhao, whose classroom teaching she frequently observed (once or twice each week) and with whom she shared her lesson designs and discussed her lesson preparation concerns, such as how to introduce new concepts and how to organise the sequence of contents.

4.4.1.2 Lesson Implementation

During the second and third semesters (see Appendix 8.1), Doris continued to implement structured, teacher controlled classroom instruction, posing numerous easy questions to the whole class (Appendix 8.1.2); she did, however, use more fact-based (52.12%) than yes/no (37.45%) questions in the third semester. In the classroom, Doris emphasised different aspects of various mathematical topics. In the second semester, she emphasised the students' memorisation of trigonometric formulae, usually by dictating the formulae and asking the class to recite them in class. In the third semester, she invested a lot of time in introducing the origins of analytic geometry to begin a new topic.

4.4.1.3 Off-class Interactions with Students

To promote students' explorations of mathematics, Doris occasionally provided interesting problems for them to solve outside of class, largely based on her master's programme learning experiences; for example, when teaching trigonometry in the second semester, she asked students to prove the *sum* and *difference* formulae for sin, cos and tan using an unlettered graph (see Figure 4.1) and was pleasantly surprised by her students' excellent proofs. Afterwards, she praised the students' efforts in class and illustrated some of their proofs; as she explained in the interview, she assigned the problem because class time was too limited and her goal was to encourage the students' autonomous efforts.

Doris said she provided interesting mathematical problems at every possible opportunity and thought it a good way to promote students' exploration of mathematics and thought development. Her students, particularly those in Doris-Class2, responded actively to her. In addition to spending much of their spare time solving and discussing the problems, they found even harder mathematical problems and discussed their solutions with Doris; in such discussions, Doris also asked students further questions, where possible. In the interviews, Doris praised these extra-curricular teacher-student interactions, describing them as a sort of 'reciprocal challenge' interaction. The interactions combined learning and teaching for her and were an enjoyable experience for both her and her students.

Figure 4.1 The unlettered
graph for the proof of
formulae

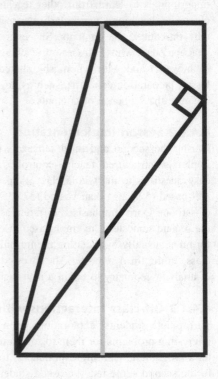

4.4.1.4 Lesson Reflection

Doris reported that she continued to reflect on the quality of her teaching almost every day and sometimes on suggestions from Teacher Zhao (her mentor) that she adopt more normative teaching behaviours[8] in class. For instance, Zhao pointed out her messy blackboard writing after observing one of Doris' teachings; Doris acknowledged the flaw and purposively improved her writing in subsequent lessons.

> For each (mathematical) example, I wrote clearly. This is also a demonstration for the students about how to rigorously write the solutions.

She reported feeling good about the improvement, stating, 'It was good when I found a structured demonstration of the whole lesson displayed on the blackboard as soon as the class ended'.[9]

In her reflections, Doris emphasised the differences in her teaching between the two classes and how such differences stemmed from her 'instinct' with respect to adjusting her teaching. She explained that, for example, 'I am used to teaching slowly in Doris-Class1 and increasing the difficulty of part of the contents or mathematical examples in Doris-Class2' when teaching the same contents. It appears that Doris's 'instinct' refers to the rules she summarised and implemented to cater to the two different classes, probably based on her knowledge of the students. She claimed, 'Doris-Class1 needs more guidance', while 'for Doris-Class2 students, they are likely to have kind of intuitive responses to my teaching'. Regarding knowledge of the content, she said that, because of her 'instinct', she sometimes had an 'inspiration' to deal with specific concepts in teaching. For example, she used the concept of *the smallest positive period of a trigonometric function* as an analogy in her definition of the angle of inclination but expressed that this 'inspiration' was contingent on her mood; 'if I am in a good mood, I will likely experience such inspiration; but if not, I cannot think of it'. Doris said that she needs more teaching experiences to turn contingency into inevitability and perform better more consistently.

---

[8] Normative teaching behaviours denote a set of rules that teachers should follow in classroom teaching, such as speaking standard Mandarin, speaking and writing in a structured way, and so on.

[9] A structured blackboard demonstration represents one of the rules in 'normative teaching behaviours'. It is considered one of the most important basic skills of teaching practice in China, along with the main content taught in that lesson and structure of the teaching process. As Doris explained, the structured demonstration reflected the coherent mathematical procedure she had conveyed to her students.

4.4.2 Beliefs Reflected in Teaching (Enacted Beliefs)

As the interviews examining teachers' stated beliefs were not carried out in the medium stages, teachers' beliefs were analysed based on their teaching activities. Doris's teaching demonstrated her conviction that mathematics was stable and that she emphasised the rigour and importance of mathematics as well as the connections between mathematical topics.

In the second semester, Doris discussed the stability of mathematics in a whole-class discussion of whether mathematics or physics was more important. The extemporaneous discussion stemmed from a comment by the class's physics teacher that 'physics is more important than mathematics'. Doris did not agree and initiated the discussion, asking students to express their ideas in a diary, as one of the assignments. Closing the discussion, Doris expressed her belief that 'mathematics is the base of all scientific subjects... mathematical knowledge is consistent with logic'; moreover, although 'the knowledge had been developed long years ago... it is the stability of mathematics that makes the classics and the truth' (Doris-R2-CO1; beginning of the lesson).

In the classroom, Doris emphasised the importance of students' acquisition of rigorous problem-solving procedures by pointing out students' misuse of the format. For example, she asked the students to dictate the rules of signs of trigonometric values of the angles in different quadrants and emphasised each step of procedure to solve the problems (e.g., the necessity of including the closing marks (Doris-R2-CO2)). In the post-class interviews, she stated that she considered these practices to be effective and necessary approaches to assessing students' learning of mathematics, because 'rigour' is an important feature of mathematics *per se*.

Doris focused on the importance of learning mathematics by comparing problems that can only be solved using mathematics to those that could be solved on a calculator; as she put it, 'mathematics is useful and beautiful'. She also emphasised students' own explorations of mathematics by providing interesting mathematical problems for them to do after class time.

In the third semester, Doris focused on students' interest in learning mathematics by integrating history and culture into her teaching. She claimed to 'always have the awareness to let [the students] know these things due to my major in HPM'; to introduce the new topic of analytic geometry, for example, she first discussed its origins (Doris-R3-CO1). In the interviews, she explained that 'even though this may not be helpful for them to understand the specific content, I just want them to know the 'thing behind', the way in which the content was situated in mathematics history as well as mathematics development'. Doris particularly highlighted the 'thing behind' mathematical facts, characterising it as the connections between facts. As she claimed in class, 'the essence of analytic geometry is using algebra as the tool to solve geometric problems', and 'analytic geometry shows the perfect combination of algebra and geometry; so does the symbolic-geometric combination'.

4.4.3 Knowledge in Teaching

Doris's teaching revealed that she involved students in experiencing the derivation process of mathematical formulae using generic and specific mathematical examples. She also emphasised the subconscious use and accumulation of knowledge in teaching and attempted to integrate mathematical history and culture into teaching.

During the second semester, Doris alternatively used generic and specific mathematical examples to involve students in the derivation process of mathematical formulae. For example, when proving some *trigonometric identities* in Doris-R2-CO1, she first gave a generic example—given that $\cos\alpha = t$, find $\cos(-\alpha)$, $\cos(\pi + \alpha)$, and $\cos(\pi - \alpha)$—and gave students time to think about the solution. When she realised the students would likely struggle to solve the problem, she provided a specific example—what is the value of $\cos(-\frac{\pi}{3})$—for which they could more easily find a feasible solution strategy (graphing) to make it easier for her to demonstrate the derivation of the *trigonometric identities* using a similar strategy. Meanwhile, she carefully illustrated the expansion of the formula's use in specific examples to its generic use in a smaller range, and finally to the real number field to demonstrate the derivation process to the students. In the post-class interviews, Doris reported that she did not consciously teach in this way but that she 'just wanted them (the students) to experience the derivation process and to experience the mathematical thinking (reasoning) and the use of the mathematical method: the symbolic-graphic combination'. She also pointed out that 'I did not deliberately do it', and explained that she might have been influenced by her own learning experiences as a upper secondary school student.

> I think the pattern I am now using for teaching probably reveals how my teacher taught me. It is not about specific knowledge but his teaching habits. I became used to thinking about how and why he taught in this way when I was in high school. Even now, I still remember some excerpts or teaching methods he used.

She again emphasised the role of her subconscious in the accumulation of relevant knowledge and her appreciation of her master's programme learning, particularly regarding the topic she had just taught (*trigonometry*). She claimed that the related major HPM readings contributed to her comprehensive understanding of the content knowledge and provided her with interesting mathematical problems that helped promote students' exploration of mathematics.

During the third semester, Doris integrated the history and culture of mathematics into the introduction of *analytic geometry*. She used her reading of Descartes' geometry (Descartes & Yuan, Trans., 2008) to introduce the origin, concept and significance of analytic geometry to students. In doing so, she emphasised that the essence of analytic geometry lies in its 'combination of algebra and geometry' and its significance in the fact that 'the combination helps in solving problems, especially the geometric problems that could not be solved in the past'. She also provided examples that could not be solved by pure algebra or geometry to demonstrate its significance.

In the post-class interviews, Doris reported that she never abandoned the idea of promoting students' interest in mathematics by introducing HPM to the class but noted that she needed 'a proper mathematical topic to use HPM', such as introducing *analytic geometry*. To integrate mathematical history into teaching, she said that she reviewed relevant historical knowledge, proudly stating that she knew how to search and review the historical knowledge she needed thanks to her master's programme, although it did not provide her with all of the requisite historical knowledge. However, 'it was not enough that I should only have the historical knowledge', Doris stated, and so she also appreciated Zhao's support, both spiritual and practical. The spiritual support Zhao provided was the consistent encouragement he gave to Doris to invest effort in integrating mathematical history and culture, particularly when she felt tired and wanted to give up. For several specific topics, he also encouraged her to consider matters from a different perspective (e.g., from the advanced mathematics). In terms of practical support, he provided suggestions based on his rich teaching experience when Doris faced quandaries in her teaching. For example, when Doris perceived conflict between the mathematical knowledge in her mind, her knowledge of her students' cognition, and the records in mathematical history or her textbook, Zhao usually gave his choice as a suggestion. Eventually, Doris summarised and devised a 'rule' to deal with such sources of confusion: 'for the man-made concepts (e.g., the equations of straight lines), the direct instruction is usually used, that is, just tell them (the students)'; while 'the indirect instruction is likely to introduce some other concepts, which have a rich history behind them and the concepts per se are quite interesting'.

These concepts usually have rigorous definitions in mathematics and are not easy for students to understand. Hence, it is necessary for the teacher to help students analyse the concepts or even teach them using a heuristic method.

4.4.4   Summary and Discussion

At the medium stage, it appeared that Doris still emphasised the importance of the teacher and the content in her teaching. Meanwhile, she also attempted to promote students' exploration of and interest in mathematics by providing interesting mathematical problems and integrating the history and culture of mathematics into her teaching. In her teaching process, she had more interactions with her colleagues and received support from them, especially her mentor, Teacher Zhao. Zhao also caused her to pay greater attention to her normative teaching behaviours in class.

4.4.4.1 Individual-level Orientation to Learning

Doris's teaching at the medium stage revealed that she emphasised the static features of mathematics and the connections between mathematical facts, which appeared to be consistent with the mixed beliefs she reported at the beginning. In particular, she integrated the history and culture of mathematics into her teaching to promote students' exploration and interest in teaching. The difference was that, in the second semester, she only used extra-curricular time to promote students' practice with interesting mathematical problems; in the third semester, she encouraged students' learning of historical materials and then integrated her historical knowledge into the introduction of new contents to increase students' attention to what she regarded as the essence of mathematics—the connections between mathematical topics and mathematical methods.

These findings indicate the reciprocal relationship between Doris's experiences, pedagogical beliefs and knowledge and teaching practice. It appears that Doris was on the way to practising the teaching she believed in by drawing on the knowledge she obtained from her learning experiences, with respect to both teaching content (e.g., historical knowledge) and new learning (e.g., searching and reviewing). Her accumulation of new knowledge to inform her teaching practice was carried out during the teaching process through her engagement with the materials, interactions with the students, and communications with other teachers. Given her students' positive responses and her mentor's support, Doris may be encouraged to continue her teaching.

However, it also appears that Doris' incessant teacher-controlled instructions, the lack of class time for discussing interesting mathematical problems, the unified teaching schedule, and the heavy load of required content constrained her promotion of students' explorations and interests. Thus, Doris's situated teaching environment also played an important role in her teaching.

4.4.4.2 School/organisational Level Orientation to Learning

The teaching environment (LPG) seemed to focus on teaching required content within a certain narrow timeframe, as demonstrated by the unified teaching schedule, students' exercises and progress tests and the collective lesson plans to which the teachers referred. The teaching materials likely conveyed several rules to Doris, such as writing her lesson plans in a formal way, and further conveyed the knowledge for teaching each lesson, particularly regarding content, the curriculum and the students. The environment also impacted Doris' teaching through interactions between her and her colleagues; Teacher Zhao urged her to adopt more normative classroom teaching behaviours and pointed out areas that required improvement, instructions that Doris apparently took to heart. At the same time, her desire to promote the students' interest by integrating mathematical history into her teaching was also encouraged and supported by Zhao in the form of mentoring as Doris practised the teaching.

4.5   The End Stage of the Two-year Professional Learning

This section first presents the main features of Doris' teaching practice and her teacher beliefs and knowledge at the end stage by analysing the data in the same way as in the beginning stage. It then describes her own reflections on her teaching and her mentor's words from the supplementary data to triangulate our interpretations of her professional learning. It is then followed by a summary and discussion of the overall learning outcomes and approaches used to achieve learning outcomes based on the findings across the entire two-year period.

4.5.1   Main Features of Doris's Teaching Practice

At the end stage, Doris relied on ready-made lesson plans that were consistent with her intended lesson designs, and did not observe other teachers' classroom teaching owing to her heavy workload. Her classroom instructions remained structured and teacher-controlled but she integrated the history and culture of mathematics into her teaching. She reduced the frequency of her post-class reflections but paid greater attention to inappropriate instructional designs and their modification.

4.5.1.1 Lesson Preparation

Doris reported that she was particularly busy with school activities (supervising Doris-Class1 students' work on school discipline[10]) and her extra teaching load (taking over a third mathematics class for an absent teacher) at that time. Therefore, she did not have time to observe other teachers' instructions or even to prepare her own teaching. To prepare her lessons quickly, she devised a teaching design, searched for and selected the appropriate lesson plans online and then made revisions.

4.5.1.2 Lesson Implementation

At the end stage, Doris continued to implement structured and teacher-controlled classroom instructions by posing large numbers of easy questions to the whole class (see Appendix 8.1.2). Compared with the third semester, she used more yes/no questions (47.13%). In the observed lessons, she significantly integrated the history and culture of mathematics into classroom teaching; specifically, she changed the foci of the lessons from teaching the formulae for the volume of solids to teaching the history of Zu Geng's principle[11] to illustrate the derivation process and the use of the formulae in the history of mathematics. She explained that her reason for this adjustment was that she believed Zu Geng's principle was more important for students to learn.

> I want them (the students) to know this. Even if one day they cannot remember the formulae of the volumes, but they know that the derivation of the formulae uses Zu Geng's principle.

4.5.1.3 Lesson Reflection

Doris stated in the interviews that she did not frequently reflect on her teaching at that time but on any mistakes she may have made in implementing the instructional plans. She also emphasised that her reflections needed foci because 'the reflection should not be so inflexible' and 'it should depend on the actual

---

[10] This task was undertaken by all Grade 10 and 11 students to involve them in the management of school discipline. With the class as the unit, the students supervise and assess other classes' discipline for one week, in turns.

[11] Zu Geng is a famous ancient Chinese mathematician. Zu Geng's principle is similar to Cavalieri's principle.

situation'. With focused reflections, she thought she could make corresponding modifications to improve her instructional design. Furthermore, when asked about the frequency of such modifications, she said 'not often, two or three times in this semester', because she 'was competent to conduct the regular teaching' that was required by the environment.

4.5.2   Doris' Teacher Beliefs

At the end stage, Doris's teacher beliefs were also analysed based on what she stated in the interview and how she behaved in teaching to develop a comprehensive understanding of her beliefs.

4.5.2.1 Stated Beliefs

From the interview (see Appendix 8.2), two types of mathematics were identified in Doris' stated pedagogical beliefs—school mathematics and scientific mathematics.

She emphasised that her teaching and learning of school mathematics was performance-oriented. Teachers must try their best to acquire the requisite teaching skills to convey knowledge clearly to students in class. Students must engage in sufficient practice to perform well in examinations, and understanding the content that is likely to be included in the examinations is essential. Doris reported that environmental pressure made the teaching of school mathematics her primary task. This left the students little time to think, despite the fact that she considered thinking important for learning scientific mathematics.

Ideally, Doris preferred teaching scientific mathematics. She believed that greater student effort was required to learn it; teachers simply need to provide the necessary assistance, promote the students' thinking and conduct proper assessment. However, Doris claimed that teaching scientific mathematics was impractical in her situated environment because the pressure surrounding the *Gaokao* obliges teachers, students and parents to focus on performance-oriented teaching. Nonetheless, Doris emphasised the importance of promoting students' interests and developing their mathematical thinking and insisted that the best way to do so was to integrate history and culture into mathematics teaching. She stated that if time permitted, she would invest greater effort in doing this in the future.

When asked how she conceptualised mathematics, Doris said that she regarded it as a rigorous system combining invention and discovery in its interactions with the real world. She thus held a mixture of traditional and non-traditional beliefs about the nature of mathematics.

4.5.2.2 Enacted Beliefs

In her teaching, Doris emphasised content, including both mathematical facts and mathematical methods and thinking. When integrating the history and culture of mathematics into her teaching, she seemed to focus more on the derivation process of formulae than on the formulae *per se*, even though derivation was not a focus of examination. She also considered mathematical methods, principles and thoughts to be the essence of mathematics. In her post-class interviews, Doris reported that she hoped her students 'could obtain as much as possible mathematical knowledge', particular mathematical methods and thinking, which would last long in their memory.

Doris also stated that the purpose of integrating history and culture into teaching was to promote students' interest in mathematics; encouraging students to recognise the beauty and usefulness of mathematics may result in their having a positive attitude towards mathematics learning. In the classroom, Doris appeared to focus on the promotion of the students' interests in addition to their performance. She emphasised that 'proper memorisation was necessary' so that students could respond swiftly to teachers' classroom instructions and was conducive to good homework and progress test performance. However, she also observed that 'it would be more meaningful if the memorisation was based on understanding'.

There appeared to be a contradiction between Doris's integration of history and culture into teaching for students' interest and her implementation of a unified teaching pace for students' performance. She believed that constant drilling and practice were unnecessary and reduced the number and length of the exercises completed by the students to facilitate the integration of mathematics history and culture into her classroom teaching yet still maintained the same teaching pace as other teachers; however, her students performed significantly more poorly than she expected on the unified homework. To help students 'stand firmly on their feet', she had to spend more time reviewing the errors they made when completing their exercises. This prompted her to question the effectiveness of her teaching, her knowledge of her students and whether she had the teaching experience required to make correct judgements. Being situated in an environment that valued students' performance so highly made it more difficult for her to implement her ideas of promoting students' interest by integrating history and culture into teaching.

4.5.3 Doris's Teacher Knowledge

4.5.3.1 'Discussing Mathematical Problems'

Compared with the beginning stage, Doris appeared more confident in discussing mathematical problems. She did not spend time solving them, and nor did she care much about the answers. While she still began by summarising the mathematical topics involved in the problems and then focused on students' acquisition of the knowledge of the contents and skills needed to work out the problems, she now also emphasised the importance of summarising and intensifying those skills for students' good *Gaokao* performance (Appendix 8.3).

By the end of the task, Doris had even reflected on the task *per se*, which she saw as a sort of teaching competition that required corresponding training. She believed that a comprehensive understanding of the curriculum's mathematical contents may help teachers to perform such tasks but would require rich teaching experience. Thus, she claimed that her lack of teaching experience prohibited her from performing well in the task.

4.5.3.2 Knowledge in Teaching

Doris's use of various types of knowledge emerged in her teaching, which integrated knowledge of both Western and Eastern mathematical history. For example, she used both Greek and ancient Chinese proofs of the formula for the volume of spheres in her teaching of Zu Geng's principle and its applications. At the same time, observation of her teaching revealed that Doris took her students' personalities into consideration when making her teaching decisions, and employed several approaches to integrate history and culture into her teaching, including illumination, modules and history-based approaches (Jankvist, 2009). Tzanakis and Arcavi (2000) asserted that students may develop a deeper awareness of history and mathematical topics through such teaching. In class, Doris carefully illustrated historical facts (e.g., mathematicians and their achievements) and expressed her preference for ancient Chinese mathematicians' work, which may also have contributed to the students' sense of nationalism and patriotism. In the post-class interviews, Doris reemphasised that the use of knowledge in teaching relied on her 'subconscious' instincts but that her past learning experiences (in particular, her learning of HPM during her master's programme) and recent teaching experiences played important roles in her accumulation and acquisition of knowledge.

4.5.4   Doris's Self-reflection on Her Learning and Words from Zhao

4.5.4.1 Doris's Self-Reflection on Her Learning

Doris's reflections on her teaching consisted of her own comments on the two-year teaching period and the factors that influenced her teaching.

Generally, Doris believed that she encountered no difficulties in completing 'the basic teaching task' of teaching mathematics to Grade 10 and 11 students—that is, she taught the curriculum contents within a regular teaching schedule, and her students performed well in unified examinations. Over the two-year teaching period, she reported that she focused entirely on content knowledge and acquiring the necessary pedagogical knowledge (e.g., whether to introduce content directly or indirectly). However, Doris emphasised that the most important benefit that she gained during the period was an understanding of 'what exactly should be done in teaching'; that is, rather than focusing too much on her students' performances in examinations, she should be teaching them the essence of mathematics. According to Doris, the essence of mathematics lies in the connections between mathematical topics, which reveals the subject's logic, beauty and usefulness; however, she also claimed she could not yet teach the essence of mathematics well, as she was still learning.

Doris believed that teachers should first consider students' demands and feedback, which is the biggest motivation for teaching. She stated that she was encouraged by the students' positive responses to her approach of integrating history and culture into her teaching and by their good performances relative to other students in the unified examinations. She particularly regarded her enjoyable interactions with students as an effective process for mathematics teaching and learning. This gave her the confidence to continue promoting students' interest in mathematics through her teaching.

Doris also appreciated the good atmosphere the school provided within which to implement her teaching. The school (particularly her LPGs) not only encouraged teachers to learn from one another by observing others' classroom instructions and freely discussing teaching but also supported teachers in practising their own teaching ideas, with relatively little pressure relating to students' performance. According to Doris, the most experienced teachers in the LPG (i.e., the group leaders) played an essential role in creating a good atmosphere.

> They did not particularly emphasise the students' performance, despite their brilliant achievements in this regard. Moreover, they always emphasised the importance of the non-utilitarian view of mathematics teaching.

Doris was particularly grateful to Teacher Zhao, her mentor and one of the two experienced teachers in the LPG, for his encouragement and support for her integration of mathematical history and culture into her teaching.

4.5.4.2 Words from Teacher Zhao (Doris's Mentor)

Based on an interview with Zhao, his stated beliefs as a mathematics teacher, his instructions for Doris's teaching as her mentor and his comments on Doris' teaching are examined.

From Zhao's perspective, mathematics forms the basis for the development of modern science and technology and has considerable real-life significance; however, most people struggle to have a comprehensive understanding of mathematics, himself included. Thus, Zhao did not say what mathematics represented for him; he simply acknowledged its importance and complexity.

Regarding mathematics teaching and learning, Zhao identified two approaches: utilitarian and non-utilitarian. The utilitarian approach pointed to the *Gaokao*'s screening function. Zhao claimed that after two decades of teaching in an environment that emphasised the *Gaokao*, he was accustomed to teaching examination-centred mathematics. To promote students' performance in the examination, he focused on the structured mathematical knowledge in the syllabus and attempted to increase 'the positive variation' in students' achievements in related examinations. For example, he usually encouraged students with sufficient knowledge to complete more exercises at higher difficulty levels and those with less knowledge to practise more basic exercises more frequently. On the other hand, Zhao's non-utilitarian view advocated romanticism and idealism in mathematics teaching and learning of mathematics, which Zhao regarded as the motivating force for the development of mathematics. However, he did not specify what he meant by romanticism and idealism.

Zhao stated that he supported Doris's beliefs in promoting students' interests and thinking in mathematics because it revealed that Doris shared his non-utilitarian view. He also indicated his interest in her approach to integrating the history and culture of mathematics into teaching. He enjoyed reading the HPM-related materials provided by Doris and discussing with her ways to integrate history into her daily teaching to promote students' interest and allow mathematical thinking to permeate their learning. He appreciated Doris' solid knowledge base and specific teaching ideas as well as her diligence and vigour in teaching. He asserted that Doris could become one of 'the most promising teachers' in the future, but noted that she still her normative teaching behaviours required some improvement, as China holds basic teaching skills to high standards, and meeting these standards was the only way Doris could become an outstanding teacher 'in the eyes of the world':

> For the blackboard writing, we require that what is presented on the blackboard should be clear, methodic and normative without any mess. For the language used in class, we hope it should be succinct, concise and effective without any nonsense words, just like academic language.

4.5.5 Summary and Discussion

At the end stage, it appeared that Doris continued to devote herself to teacher-controlled classroom instruction but also emphasised the integration of mathematics history and culture into her teaching to promote students' appreciation of mathematical methods and the thinking involved in the derivation of mathematical formulae. In her stated beliefs, Doris distinguished the teaching of school mathematics from the teaching of scientific mathematics; the former was teacher-centred and performance-oriented, while the latter was learner-centred. Although Doris considered teaching scientific mathematics to be her ideal teaching approach, environmental pressures obliged her to focus on teaching school mathematics. In discussing mathematical problems, she focused on students' skills in working out problems that students were likely to encounter in the *Gaokao*; however, she continued to emphasise the importance of promoting students' interests and mathematical thinking and considered the integration of mathematical history an effective teaching approach.

4.6 Doris's Professional Learning

From the above descriptions of Doris' teaching over the two years, both fixed and changed aspects were detected in her teaching practice, teacher knowledge and beliefs, which constituted the overall learning outcomes of her teaching.

4.6.1 Overall Learning Outcomes

Across the two years, Doris consistently held a mix of traditional and non-traditional beliefs about the nature of mathematics and emphasised both the rigorous contents of mathematics and the connections between mathematical topics, which she regarded as the essence of mathematics. By the end stage, she specifically differentiated between two types of mathematics—school mathematics and scientific mathematics—and their correspondingly different teaching and learning requirements. Doris considered teaching for a scientific mathematics subject, which is learner-centred and requires proper assessment that differs from the *Gaokao*, to be her ideal form of teaching. By contrast, teaching for school mathematics—the teaching she was forced to implement due to environmental pressures—is performance-oriented and teacher-centred. Doris consistently believed in promoting students' interest in mathematics by integrating the history and culture of mathematics into her teaching, an idea that seems to have been enhanced by the end stage owing to the effectiveness of her practice.

Doris consistently practised the teaching of school mathematics, which she considered her 'basic teaching task' and which emphasised the teacher, the content and students' performance in public examinations. Over the two years, Doris relied on teacher-controlled classroom instructions, followed the unified teaching schedule and kept pace with other teachers. To implement this sort of teaching, Doris had to rely on the available teaching materials to form a comprehensive understanding of the curriculum contents, students' characteristics with respect to learning specific content, and formal ways to write lesson plans. Meanwhile, she also needed to develop the normative teaching behaviours advocated by her situated environment. Doris also implemented practices aimed at promoting students' interest in and exploration of mathematics—and thereby their mathematical thinking—by integrating mathematical history into her teaching. Although she claimed that the integration of history into teaching relied extensively on the mathematical contents *per se*, the historical facts were gradually introduced and integrated into her teaching over the two years, appearing first in extra-curricular activities/homework and then in some classroom activities (e.g., introducing new topics) before being fully adopted in the class.

Doris' teaching and her own reflections on her learning of teaching reveal that she had acquired and accumulated knowledge of the content (including curriculum contents and the history of mathematics), knowledge of students, and knowledge of teaching (for both 'the basic teaching task' and the integration of mathematical history) through the integrated use of these various categories of knowledge in her teaching practice.

4.6.2 The Ways to Achieve Her Learning Outcomes

First, it appears that the experiences that Doris brought to her teaching in School A played an important role in her two-year professional learning experience—in particular, her teacher beliefs and knowledge. For example, her specific learning experience of HPM from her master's programme provided her with the idea of and necessary knowledge for promoting students' interest in mathematics by integrating the history and culture of mathematics into teaching. The one year's teaching experience and instruction she received as a student contributed to her perceptions and knowledge of 'the basic teaching task'.

This suggests that Doris's past experiences led to her independent implementation of teacher-centred, content-focused teaching at the beginning stage. She was able to use the available teaching materials to prepare lessons independently and to adjust her teaching to reflect students' feedback and the school-developed teaching material. Her new learning supported her teaching but was focused more on her knowledge of the content and of the students. Meanwhile, the students' good performance on the unified progress test caused Doris to recognise the effectiveness of the teaching she had implemented, which may have reinforced her beliefs in relation to school mathematics teaching. Her accumulated knowledge and enhanced beliefs helped Doris to continue her teaching practice, and implementing her teaching allowed her to accumulate new knowledge and enhance her beliefs. The reciprocal influences among Doris' experiences, teacher beliefs, knowledge and teaching practice constituted her professional learning for school mathematics teaching.

However, based on Doris's statements in post-class interviews, it appears to have been difficult for her to detect precisely when and how the influences occurred; she could only give some specific examples within the teaching context. For example, she said that her teaching of a specific topic was inspired by an indistinct memory of having been taught similarly, but she could not offer specific details.

Doris seemed relatively clear about the influences of her situated teaching environment. In the beginning stage, she endeavoured to help her students 'stand firmly on their feet' as they competed with their peers in the unified progress test. Later, she clearly pointed out that environmental pressure forced her to implement performance-oriented school mathematics teaching. As several researchers have recognised, the pressures relating to large-scale public examinations cause Chinese mathematics teachers to adopt more traditional teaching practices (e.g., Ma, Lam, & Wong, 2006). This was consistent with Doris's mentor's stated belief in collective beliefs regarding trust (in Gaokao's screening effect) and efficacy (e.g., teacher-centred and content-focused teaching is effective for students' good Gaokao achievements). This revealed the shared norms of the environment in which they conduct their teaching and which will govern their teaching behaviours and impact their pedagogical beliefs and knowledge (Tschannen-Moran, Salloum, & Goddard, 2014).

The above descriptions of Doris teaching suggest that the environment impacted her professional learning via various means, such as the provision of teaching materials, collective teaching activities (e.g., unified progress tests) and one-to-one mentoring. Her teaching practice, teacher beliefs and knowledge may all have been influenced. For example, she implemented the school mathematics teaching under the constraint of the unified teaching schedule, with the teaching pace and assessments arranged by her LPG, and adopted more normative teaching behaviours to suit her mentor's requirements. In her teaching practice, she also accumulated related knowledge by studying the teaching materials provided by the environment, which were centred on the centralised curriculum and the public examination (e.g., Cai, 2012; Zhang, 2012). Through her interactions with other teachers, it is also likely that Doris's beliefs were influenced by the collective beliefs held by other teachers in her community (i.e., the LPGs). Although it is difficult to determine precisely how the environment influenced specific elements of practice, belief or knowledge, its influences on each element in the teaching subsystem focused Doris's professional learning on teaching for school mathematics. Combining the interactions within the teacher's individual elements, Figure 4.2 illustrates all influences on Doris' professional learning for teaching.

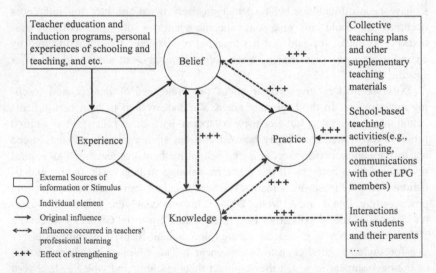

Figure 4.2 Doris' professional learning focused on the teaching of school mathematics

Doris's efforts to integrating mathematics history and culture into her teaching were also evident across the two years. Although she did not clearly state that this was her ideal form of teaching, her mentor, Teacher Zhao, seemed to verify that it was consistent with the non-utilitarian teaching style they preferred to adopt. Doris aimed to promote students' interest in and exploration of mathematics as well as the development of their mathematical thinking; descriptions of her teaching revealed that her professional learning for the integration of mathematical history (Figure 4.3) adhered to the same model as her learning for the teaching of school mathematics (Figure 4.2). However, her specific teaching practices required corresponding teacher education support during her pre- and in-service periods. With her master's programme HPM learning experience, Doris was able to bring relevant pedagogical beliefs and knowledge to bear on her teaching. As Doris stated, the school did not exert excessive pressure on her with respect to students' performance, which allowed her to implement her own teaching ideas. In particular, her mentor provided encouragement and support for her teaching.

Given Zhao's non-utilitarian approach to mathematics teaching, it appears that experienced mathematics teachers in China may also hold beliefs about teaching that differ considerably from those that focus on students' *Gaokao* performance. This is consistent with Chen and Leung's (2014) findings, based on their review of previous studies, that Chinese teachers realise the importance of 'student-centred' teaching. In addition, the High School Mathematics Curriculum Standard (experiment) issued by China's Ministry of Education (2003) deemed 'reflecting the cultural value of mathematics' to be a fundamental idea and suggested the integration of mathematical culture into the upper secondary school mathematics curriculum. It appears that Doris's relatively student-centred teaching ideas (i.e., promoting students' interests and exploration) and her integration of mathematical history into her teaching were not unique to her. However, Zhao considered the non-utilitarian approach to be idealistic and romantic, and both Zhao and Doris admitted that the *Gaokao* occupies a powerful position. Doris also highlighted the difficulties in teaching integrated mathematical history (e.g., lack of class time, students' poor performance in unified exercises); nonetheless, she tried her best to promote the students' interest by integrating mathematical history into her teaching. However, she was constrained by the unified teaching schedule and pace.

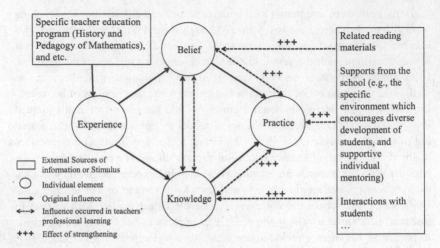

Figure 4.3 Doris' professional learning focused on the integration of the history and culture of mathematics into teaching

The Case of Jerry

5

5.1 Introduction

This chapter presents the professional learning of Jerry (pseudonym), who served in School B, a key upper secondary school located in District II. Jerry obtained a bachelor's degree in mathematics and applied mathematics from a normal university and subsequently completed a master's degree in mathematics at a prestigious comprehensive university. He reported in the interview that he did not experience many teaching-related activities during his learning in the bachelor's and master's programme. With the exception of the practicum and internship[1] that he experienced as a student teacher in the normal university, he had experienced private tutoring in mathematics and substitute teaching on a mathematics course for international students. During his two years of teaching in School B, he gradually strengthened his focus on students' performance in the large-scale public examination (*Gaokao*). Thus, his learning with respect to teaching was centred on this goal, which was significantly influenced by pressures and support from the school.

---

[1] These two activities have been explained in Chapter 4: footnote 1.

Supplementary Information The online version contains supplementary material available at (https://doi.org/10.1007/978-3-658-37236-1_5).

5.2   Background

5.2.1   Jerry's Background

Jerry reported that the teaching experience he accumulated during his four-year bachelor's programme—in particular, his internship—had changed his perceptions of teaching, especially the transformation from the role of a student to that of a teacher. He claimed that it was necessary for a teacher to develop new interpretations of the content that differed from the students' and to acquire teaching strategies to promote students' learning.

> Teaching is quite different from learning. For example, as a student, you just need to know the procedure to solve a specific mathematical problem. As a teacher, you have to teach the students not only how to solve the specific problems but also the general problem-solving strategies. Even more, you need to know how to mobilise students' thinking.

However, Jerry considered his 18 months of substitute teaching to be useless; because it involved a different curriculum for international classes, Jerry believed that the experience did not inform his delivery of the regular mathematics curriculum for Shanghai upper secondary school students.

Jerry also experienced a one-semester internship in School B after the school agreed to employ him. During that period, he observed the Grade 10 lesson preparation group (LPG) leader's teaching and helped him mark the students' homework. These activities made him more familiar with daily teaching and the students in School B, as well as the content to be taught. Jerry reported that 'the students were not as good as I expected for students in a key upper secondary school'.

Regarding his reasons for being a teacher, Jerry talked about his attraction to mathematics and to teaching as a career. He stated, 'I myself am interested in mathematics, so it is also vital for me to teach what I have learned about mathematics to students'; he also mentioned a family connection, saying, 'From my uncle's life as a mathematics teacher in a middle school, it is not a bad idea for me to be a mathematics teacher'.

5.2.2   School Information

School B is a key upper secondary school with a history of more than 50 years in Shanghai. It constitutes three independent departments: the regular secondary school department, the international department and the Xinjiang department. Jerry worked in the regular department, which mainly recruits students who intend to enter colleges or universities for further study through the *Gaokao*. The department consists of Grades 10, 11 and 12, with eight classes in each grade. One of the classes, informally called the Science Class, recruits students with better achievements in the entrance examination (*Zhongkao*), while the other classes are parallel classes. Jerry taught the same parallel class (Jerry-Class) during the two years from Grade 10 to Grade 11.

Jerry was involved in two lesson preparation groups (Jerry-LPG1 and Jerry-LPG2) across the two years. According to Jerry, owing to the two group leaders' different leadership styles, the situations of the two groups were quite different. Qian, the leader of Jerry-LPG1, seldom organised group activities in the first year, so the seven or eight teacher group members, who were divided into several small offices, had limited opportunities to discuss teaching with each other. However, they all maintained the same teaching pace and joined the same whole-grade progress tests so that their students could compete with one another. The group leader in LPG-Jerry2 during the second year was Teacher Sun. He requested that the entire grade use the school-developed exercise book for the unified students' homework, organised weekly group meetings to unify the teaching pace and shared the task of preparing the collective lesson plans among all the members. Jerry believed that these efforts saved him much teaching preparation time and thereby made his teaching easier. He admitted that the activities were rather formalised, as they did not provide many opportunities for the group members to learn from one another by communicating or discussing matters; however, he also said that the situation was too 'complicated' to change.

> Aside from myself, the other teachers are experienced teachers, and most of them have served as group leaders or even administrative staff. It is not easy for them to obey others' suggestions, even the leader's.

Qian and Sun, respectively, were assigned as Jerry's mentors during the two years and influenced his teaching in different ways owing to their different teaching styles. According to Jerry's descriptions, Qian focused more on the independent mathematical knowledge in each lesson and students' understanding of the knowledge through clear explanations. Sun emphasised the connections between mathematical topics for students' comprehensive understanding of the knowledge, since the *Gaokao* particularly requires integrated use of content.

5.3   The Initial Stage of Two-year Professional Learning

This section presents the teaching practice implemented by Jerry and the teacher beliefs and knowledge he held at the very beginning of the two years.

5.3.1   Main Features of Jerry's Teaching Practice

Jerry attempted to prepare lessons independently before starting teaching by referring to the textbook and other supplementary teaching materials available in the book market, and he planned to use the courseware (e.g., PowerPoint slides) for the implementation of each lesson he prepared. However, he did not implement the classroom teaching as planned. Instead, he observed (almost daily—that is, four out of six lessons per week) and emulated his mentor Teacher Qian's classroom instruction and used more blackboard presentation. He implemented structured, teacher-controlled instruction in class and conveyed his understanding of the content in detail to the students, with emphasis on memorising rules to complete exercises quickly. After class, he reflected in detail on his teaching behaviours as well as his improvements and recognised his weaknesses in terms of coherent teaching procedures when compared with Qian's teaching. He assessed his teaching based on the students' performance in progress tests and tended to promote students' achievements by encouraging them to work hard. Jerry also appreciated Qian's specific guidance for his classroom teaching, though it was not frequently offered. In most instances, mentoring was a unilateral arrangement, with Jerry learning from Qian by observing his classroom teaching and initiating questions.

5.3.1.1 Lesson Preparation
Jerry had prepared the plans for several lessons during the summer break by referring to the textbook, other teaching materials he had bought in the book market and sources downloaded from the Internet. He reported in the interview that he wrote the teaching plans in the form of courseware (PowerPoint slides) and planned to use them directly in classroom instruction from the beginning. However, he used the blackboard instead, since he thought that 'the contents would be visible for longer on the blackboard for students to memorise'.

Another significant phenomenon found in Jerry's teaching was that he observed Qian's classroom teaching and implemented Qian's teaching approach. Jerry did not mention this in the interview when asked how he prepared his lessons; rather, it was recognised in the post-class interviews when he was asked why he emphasised the use of scripts[2] in memorising rules to complete exercises. For instance, when teaching the basic properties of inequalities, he emphasised the summary of the scripts as important for memorising the properties, such as '同向不等式, 能加不能减', which means that for maintaining the direction of the sum of two inequalities with the same direction, the two could do the addition but not the subtraction. Jerry admitted that he had learned it from Qian and stated that he trusted Qian's way of teaching and sought to emulate Qian's classroom instruction, particularly when introducing important new content. His observation of Qian's teaching provided him with resources, such as the mathematical examples Qian used in class and the ways in which he introduced specific content and highlighted key points. Thus, Jerry said that he observed Qian's classroom instruction almost every day and asked him questions based on his observations. During the data collection period, however, discussions between Jerry and Qian were seldom observed.

5.3.1.2 Lesson Implementation

Jerry implemented structured classroom instruction, typically starting lessons with a review of problems students had encountered with the previous day's homework and previous knowledge relating to the new content, followed by the introduction of more new content based on connections between different aspects of mathematical knowledge (Appendix 9.1.1). Jerry conveyed his understanding of the new content to students in detail and emphasised the specific scripts that students could use to memorise the rules for doing exercises. Jerry also held the authority to initiate questions in class and determine the correctness of answers; the majority of questions were yes/no questions (59%), instructing questions (62.15%) or non-response questions (55.66%) (for details, see Appendix 9.1.2), indicating that his classroom instruction was teacher-controlled.

5.3.1.3 Lesson Reflection

At the beginning stage, Jerry typically reflected on his teaching after each class. He focused in detail on his behaviours in class, such as whether he erased the

---

[2] Scripts such as '同向不等式, 能加不能减', as used by Jerry and Tommy in their teaching, are typically catchy phrases in Chinese that the teachers summarise for students to easily memorise.

words on the blackboard too soon or over-explained a concept. After several days, he believed that his teaching was improving, particularly in terms of 'controlling the pace of the classroom teaching'. He had learned about the appropriate capacity (see footnote 3 in Chapter 4) of his lessons and the importance of knowing his students well so that he could ask appropriate questions to ensure that his teaching progressed smoothly. Nonetheless, Jerry believed that he still had much to learn from Qian in terms of making his classroom instruction more coherent. To his mind, coherence relied not only on the use of language movers but also on the content's logic. In this sense, he did not believe that he had wholly grasped Qian's approach to implementing his instruction.

Jerry particularly mentioned the modifications he made based on Qian's instruction. He referred to these imitations as 'innovations' in his teaching; for instance, he mentioned two 'innovations' in Jerry-R1-CO3: one was an adjustment to when to use a mathematical example, and the other was a different way to introduce a property of inequality ('if $a > b > 0$, then $a^n > b^n$, where n is a positive integer'). He claimed that rather than directly telling students the property as Qian did, he introduced them to the derivation procedure.

> I proved it out with the second property. I asked them (the students), firstly, the easier one $a^2 > b^2$, and then gradually increased the value of n. See, I proved it gradually, while he (Qian) just gave out the formula.

In doing so, Jerry 'hoped the students could experience the process of mathematical generalisation' and believed that it was 'useful for students' memorisation of the property'.

In addition, Jerry regarded students' performances in the whole-grade unified tests as an essential criterion for assessing his teaching and was not satisfied that his students had average test scores. He reflected on his teaching, noting that 'although I had already given a clear demonstration of each topic, the teaching effectiveness did not reach my expectations', and 'the biggest problem was that I cannot guarantee each student could understand what I taught'. He said that he would use more teacher–student conferences after class to supervise students' academic progress and encourage them to work harder.

5.3.1.4 Qian's Interventions in Jerry's Teaching
In most instances, Qian's mentoring took the form of Jerry actively learning by observing Qian's classroom instruction and asking Qian's advice in the way of organising classroom teaching. The induction programme required novice teachers' mentors to observe several of their mentees' lessons each semester and,

fortunately, Qian observed Jerry teaching one lesson during the investigation. Before the class, Qian notified Jerry of his coming to observe the classroom instruction so that Jerry could prepare in advance. Qian reviewed Jerry's lesson plan and offered suggestions, such as noting that Jerry's neglect of his students' cognitive level resulted in his plan being overly simple. This led Jerry to recognise the key point of that lesson and focus on how Qian dealt with it when he observed Qian's teaching of that lesson. After Jerry's class, Qian again offered the criticism that Jerry had not highlighted the key point but did not provide any specific guidance to help him improve. Jerry then reflected on the teaching by himself and admitted that he was too focused on the smoothness of the demonstration and made too few efforts to draw students' attention to the key point.

Jerry stated in the interview that such interventions were useful in showing him the lessons' foci, as was observation of Qian's teaching. Thus, he learned from Qian, even though Qian did not give him detailed guidance. However, these interventions did not frequently occur—'once or twice per week or even per month when he (Qian) was busy'. Jerry said. Jerry also recognised that he had to rely on himself in his learning and appreciated that the mentoring system allowed him to learn from Qian by opening Qian's class to him for observation and allowing him to ask Qian's advice with respect to teaching.

5.3.2 Jerry's Teacher Beliefs

5.3.2.1 Stated Beliefs

Jerry appeared to hold a blend of traditional and non-traditional beliefs about the nature of mathematics at the beginning stage, as he claimed that the mathematics he taught represented a set of rules that had always been there (see Appendix 9.2). He conceptualises mathematics as the understanding and abstraction of real-life problems and the system that can be applied to solve those problems. Understanding and abstraction took not only the form of mathematical facts but also of mathematical thinking. He considered mathematical thinking the most important element in solving countless problems and summarisation and generalisation to be important components thereof. Furthermore, he argued that mathematics is established and developed based on real-life generalisations.

Correspondingly, Jerry believed that the learning outcomes and teaching goals for mathematics teaching and learning should be students' understanding of the concepts and the development of their mathematical thinking, respectively. However, Jerry also highlighted the essential role that students' performance in examinations played in assessing his teaching. To achieve good student performance, Jerry believed in 'guiding students' to experience the derivation of mathematics and thus develop a deep understanding of the concepts and specific skills needed to solve commonly examined problems rather than 'dogmatic teaching', which focused on the memorisation of utilitarian rules and facts. As a teacher, he thought that he himself should acquire richer, advanced professional knowledge and thinking to better understand the teaching content and the students.

5.3.2.2 Enacted Beliefs

From Jerry's teaching, it seemed that his enacted beliefs focused on the contents and students' understanding, since he generally explained the contents in various ways to ensure students' understanding. The teaching he conducted was teacher-centred, with teacher-controlled classroom instruction. Based on his willing imitation of Qian's teaching, it appeared that he was used to obeying the authority of experts. In addition, inconsistencies seemed to exist between what Jerry stated and how he behaved. For example, although he was opposed to memorising rules and facts in 'dogmatic teaching', he still advocated students' use of Qian's scripts for memorising the rules for exercises; when he adopted blackboard presentations, he emphasised that their purpose was to help students memorise the lesson contents; when his detailed explanation of the contents did not result in students' good performance in examinations, he focused on supervising students' practice; he spent considerable class time reviewing the errors students made in doing homework, initiated conference with individual students after class to criticise their attitudes and work habits and encourage them practice more. These all resulted in Jerry having a heavy workload, and thereby led to his negative attitude toward teaching mathematics, which he reported in a post-class interview 'was not the teaching I expected'.

> I thought teachers just need to provide students with the thinking and inspiration to learn by themselves. I did not expect that I should do so much work. It is too tough.

5.3.3   Jerry's Teacher Knowledge

5.3.3.1 'Discussing Mathematical Problems'

When Jerry was asked to discuss mathematical problems (Appendix 9.3), he first reviewed the mathematical topics involved in the problems from a student's perspective, including any relevant knowledge they may have and what mistakes they may make in trying to solve the problems. He then illustrated his approach to teaching problems, which focused on a detailed demonstration of the solution process. Jerry believed that students 'might never know the way to solve [the problems] if [the teacher] did not clearly tell them'. Thus, he used the connections between mathematical topics and posed easy questions so that students could follow his demonstration of the solution.

5.3.3.2 Knowledge in Teaching

Jerry appeared to rely heavily on his clear and detailed explanation of the content to promote students' understanding of it. As reported in the post-class interviews, this required that he understand the relevant mathematical knowledge before the class. Jerry had confidence in his mathematical knowledge; as he said, his good performance in learning elementary mathematics led him to major in mathematics at university, while both his bachelor's and master's programmes gave him advanced mathematical knowledge. However, simply having mathematical knowledge is not sufficient for teaching; he still had to study the teaching materials and consider how to teach the content. He connected new content to other mathematical topics or their real-life applications to structure their introduction, claiming that this made his teaching more coherent.

According to Jerry, he knew what should be taught and how, from observing Qian's teaching of the same lessons. Thus, he simply needed to focus on and draw on his understanding of the content and the connections to ensure a clear, smooth demonstration. However, he also found that he knew too little about the key points of each lesson and how to highlight those points. He thought that Qian did a better job of teaching the same lessons and expressed a willingness to improve by studying related teaching materials, rethinking Qian's instruction, and considering his students' personalities.

5.3.4 Summary and Discussion

In the beginning stage, Jerry observed Qian's classroom teaching and followed his approach to teaching the content in his own class using teacher-controlled instruction. He reflected in detail on problems in his teaching (e.g., whether the explanation of a certain content was too detailed) and made modifications based on Qian's teaching. He also focused significantly on students' performance, which he regarded as an essential measure of his teaching quality. The beliefs reflected in his teaching focused on content, students' understanding, and students' performance in examinations. He also used his understanding of the contents and their connections with other elements and the real world to ensure clear and coherent instruction in class.

In his stated beliefs, Jerry emphasised content but with a particular emphasis on the importance of mathematical thinking. He even mentioned the interactions between mathematics and the real world and how it informed the development of mathematics, suggesting that he held somewhat non-traditional beliefs. However, he believed that the mathematics for teaching was defined by the curriculum and focused on students' performance as the key criterion for judging his teaching quality. In discussing mathematical problems, Jerry also focused on teachers' clear demonstration of problem-solving procedures with an emphasis on knowledge of content, students and pedagogy.

5.3.4.1 The Individual-level Orientation to Learning

The section that follows discusses interactions among the individual aspects of Jerry's teaching, including experience, teacher beliefs, knowledge and teaching practice.

The influences of experiences

Jerry brought the teaching-related experiences obtained during his university teacher education, his internship at School B and his tutorial and substitute teaching of mathematics to bear on his teaching practice. These experiences influenced his perceptions of teaching—for example, a teacher's perspective on problem-solving differs considerably from that of a student, and different curricula may be delivered in different ways. His experiences also afforded him knowledge of both subject content and pedagogical content and a profound knowledge of mathematics and pedagogy. Both his beliefs and knowledge affected his teaching practice at the beginning stage.

The influences of teacher beliefs and knowledge on teaching practice

Jerry's stated beliefs suggest that he held a static view of mathematics with respect to teaching and learning but with an emphasis on students' understanding of the content, development of mathematical thinking and good examination performance. His beliefs were reflected in his teaching, in which he used teacher-controlled instruction, followed his mentor's teaching, attempted to ensure students' understanding of content through detailed explanations, and drew on students' test achievements to assess his teaching. To achieve clear and smooth classroom teaching, Jerry relied on his understanding of the content, which he had deepened by considering the subject's internal and external connections and studying related teaching materials (e.g., the textbook and supplementary teaching materials).

However, although Jerry emphasised the importance of mathematical thinking and expressed a willingness to develop students' thinking in terms of mathematics in his stated beliefs, he implemented few measures aimed at achieving these goals, with the exception of emphasising his own clear, coherent content demonstrations. He appeared to have no specific method for promoting students' thinking and emphasised students' passive acceptance of mathematical procedures when discussing mathematical problems. Perhaps, as Fennema and Franke (1992) suggested, constraints in his teacher knowledge led to inconsistency between his beliefs and teaching practice. When he found students' unified test performance to be poorer than he had expected, leading him to reassess his teaching effectiveness, he resorted to the teaching strategy of compelling students to practise more rather than attempting to stimulate their thinking. This implies that his beliefs were focused on students' performance, indicating that traditional teaching was an important aspect of his teaching practice (Stipek, Salmon, & MacGyvers, 2001).

The above discussion demonstrates that Jerry's teaching practice was influenced by both his beliefs and his knowledge. Meanwhile, his learning was also influenced by his teaching practice, such as studying teaching materials to ensure clear, coherent teacher demonstrations. Thus, interactions took place between his beliefs and knowledge and were mediated by his teaching practice. As suggested by Ernest (1989), teachers' beliefs work as a regulator between their knowledge and behaviour. In Jerry's case, it was the interactions and intersections between these individual elements that constituted his individual orientation towards learning (Opfer & Pedder, 2011).

5.3.4.2 The School/organisational Level Orientation to Learning

According to Jerry, the school influenced his teaching before he began teaching through his one-semester internship experience, which taught him about daily teaching activities. The school then imposed constraints on his teaching with its unified teaching pace, students' homework and progress examinations. These could also be regarded as supports that Jerry, as a novice teacher, could use to learn how to arrange teaching schedules and assess students' learning. However, Jerry appears to have been primarily influenced by the pressure of the whole-grade unified tests, given that he emphasised his students' performance in the competition so much in his teaching.

The environment also supported Jerry with one-to-one mentoring, allowing him to observe and imitate his experienced mentor's (Teacher Qian's) teaching.

Qian offered advice on Jerry's teaching when necessary and guided Jerry's classroom instruction several times by reviewing his lesson plans and observing and commenting on his classroom teaching. It appears that his mentoring impacted Jerry's teaching practice and knowledge and likely also influenced his pedagogical beliefs. For example, by observing Qian, he learned the content to be taught, and by imitating Qian, he learned how to teach them; with Qian's guidance, Jerry learned about the specific teaching approaches for particular lessons (e.g., the key points for teaching the lesson). Meanwhile, his reliance on Qian's authority as an experienced teacher may have harmonised his own beliefs with Qian's, as when he adopted Qian's scripts for memorising rules and emphasised his students' memorising of these scripts to ensure their good performance. In this sense, mentoring may also have constrained the free development of Jerry's teaching. Aside from mentoring, several school-based collective teaching activities enabled Jerry to communicate with and learn from his colleagues.

5.4   The Medium Stages of the Two-year Professional Learning

5.4.1   Main Features of Teaching Practice

In the second semester, Jerry still referred to Qian's classroom instruction to guide his introduction of new concepts and focused on proper content difficulty levels and students' errors in completing exercises to facilitate their good test performance. He also implemented structured, teacher-controlled instruction in class, but interacted more with individual students and strengthened students' exercises in class. In his after-class reflections, he recognised the importance of 'managing students' in enhancing his teaching effectiveness, based on his observation of Qian's classroom teaching, and valued Qian's clear, logical and simple instructions.

During the third semester, Jerry used the collective lesson plans developed by Jerry-LPG2; his preparation relied on the instructional designs in the textbook and was supplemented with mathematical examples chosen from online lesson plans or examination papers or from students' exercise books. The atmosphere in Jerry-LPG2 differed from that in Jerry-LPG1, because the group leader, Teacher Sun, emphasised students' *Gaokao* performance in Jerry-LPG2. In this situation, Jerry explicitly stated his examination-oriented teaching orientation. He focused on students' practice in class, particularly with regard to the types of questions that frequently occurred in examinations. He also spent extensive extra-curricular time tutoring individual students who had not performed well in their regular homework and tests but reflected less often on his teaching, which he considered an effective way to promote his teaching practice.

5.4.1.1 Lesson Preparation

In the second semester, to improve students' performances in regular assignments and tests, Jerry first judged which content should be prioritised when preparing each lesson, based on test requirements and feedback from students' exercises on related topics, and then selected the appropriate mathematical examples to introduce the content in question. He also referred to Qian's classroom teaching when selecting mathematical examples or teaching procedures for key curriculum concepts (e.g. function). He still frequently observed Qian's classes and valued Qian's clear and logical scaffolding of students' learning of key concepts highly.

In the third semester, Jerry participated in collective lesson preparation organised by the new LPG leader, Teacher Sun. Sun initiated weekly meetings for all group members and assigned each member the task of preparing a different lesson. The prepared 'collective lesson plans' were then sent to all other members by the designer. Jerry stated that he implemented his teaching based on the collective lesson plans. Thus, his lesson preparation was limited to understanding the contents and modifying contents that were significantly inappropriate for his students. He admitted that, rather than providing a learning opportunity, the collective lesson plans brought convenience to his teaching, because he did not need to invest excessive time in preparing his lessons. The LPG activities included no discussions on lesson planning or reflections on the plans' implementation. As such, Jerry had no opportunities to learn from other teachers, except for Sun, whose classes he could observe and to whom he could pose questions. Coincidentally, the lessons that Jerry taught and that were observed by the author during this semester were also lessons that he had been tasked with preparing. He stated that he abided by the instructional design laid out in the textbook and selected mathematical examples with varying levels of difficulty from online lesson plans, examination papers and student exercises book to consolidate students' understanding of the contents and practice their application.

5.4.1.2 Lesson Implementation

Jerry appeared to implement his classroom instruction much as he had during the beginning stage. His instruction was structured, beginning with a review of knowledge, followed by the introduction of new content connected with that review; his instruction was also teacher-controlled, as he exercised his authority to initiate numerous easy questions and to decide the correctness of any responses (see Appendix 9.1.2). However, he posed more questions to individual students as a means of inviting them to interact with him. During the second semester, his review of students' procedures for solving problems relied extensively on their correct imitation of his demonstrated procedures, which he considered an effective way for students to acquire the solution procedures, 'because Chinese [students are] good at imitation'. During the third semester, he focused particularly on reviewing students' homework errors and the practice and summary of question types, particularly those typically encountered in examinations, to promote students' performance. In the third semester, Jerry considered it important that his teaching reflect examination requirements.

5.4.1.3 Lesson Reflection

During the second semester, Jerry compared his teaching with Qian's in his reflections. He reported that he admired Qian's ability to teach contents clearly, simply and logically and to manage students. He also noted that he wished to make efforts in these aspects himself. To this end, he focused on Qian's demonstration of new content when observing his classroom teaching. In particular, he frequently asked students to come to his office after class to complete exercises under his supervision; he believed this could help students develop good learning behaviours. As he stated, the appropriate management of students both in and out of class was beneficial to his teaching efforts. This perspective emphasised teachers' authority and student discipline, so that students could concentrate intensively on their teachers' instruction.

During the third semester, Jerry believed that his teaching should be more examination-oriented, reflecting the requirements of Jerry-LPG2, particularly its new leader. Sun had several years' experience of teaching Grade 12 students to prepare them for the *Gaokao* before he took over the LPG; thus, Sun emphasised advance *Gaokao* preparation. He focused particularly on students' acquisition of the contents and the solutions for question types required by the examination, and closely observe student progress and test performance across the entire grade to promote learning for the examination. This placed Jerry under considerable pressure, leading him to spend more time pushing students to practice more and tutoring students individually after class. Jerry complained that his increased workload made him feel tired, although the LPG provided collective lesson plans and unified question banks for him to use. Thus, he had little time to reflect on his daily teaching.

5.4.2 Beliefs Reflected in Teaching (Enacted Beliefs)

Jerry's teaching in the second and third semesters revealed his inclination towards traditional beliefs about teaching. He focused more on teaching the mathematical concepts and procedures required for examinations and student management with the aim of promoting student performance.

During the second semester, Jerry particularly emphasised the teaching of key curriculum content, repeatedly discussing them in his classroom teaching and focusing on students' memorisation of relevant mathematical formulae and their exercise practice. He also stressed the importance of students' imitation of his demonstrated problem-solving procedures for the question types required in examinations; he considered this a way for students to acquire the rigorous, standard mathematical procedures that would benefit them in tests. When introducing content and demonstrating problem-solving procedures, Jerry used numerous easy questions to ensure that the entire class followed his instruction. He also highlighted his authority in class by emphasising student discipline. According to Jerry, his class management efforts aimed to ensure the students' obedience and ensure examination performance.

During the third semester, Jerry increasingly emphasised *Gaokao*-oriented teaching, carefully tailoring his teaching to deliver content that was particularly pertinent to the examination (e.g. the equations of straight lines) and providing descriptive explanations thereof; he did not devote much class time to illustrating less-important content (e.g. the program graph). Jerry emphasised that students should practice important examination-related question types and urged them to apply the procedures he demonstrated to complete the related exercises. He called this strategy '*yi-hu-lu-hua-piao* 依葫芦画瓢' (drawing a dipper with a gourd as a model), a phrase he mentioned several times in his classroom teaching. Because the *Gaokao* usually emphasised comprehensive mathematical problems (i.e., problems involving various mathematical topics or methods for their solution), Jerry encouraged students to implement relevant practices to develop their problem-solving skills.

> You (the students) should complete more exercises for the difficult problems, especially the previous *Gaokao* problems related to the mathematical topics we are just learning. These problems are all comprehensive problems required by *Gaokao*. You should experience more such practice to acquire the skills required to solve the problems in the *Gaokao*. It is better to prepare for the *Gaokao* as early as possible.

Regarding his students' overall achievements, Jerry focused his supervision on students whose performance was below average by initiating frequent individual teacher–student conferences to correct their mistakes and push them to practise more to consolidate their content learning.

5.4.3 Knowledge in Teaching

The knowledge that Jerry enacted in his teaching appeared to focus on the examination, such as summarising teaching patterns (e.g., how to choose and use mathematical examples). He also focused on preparation for the examination, distinguishing between 'important' and 'unimportant' content and question types, based on examination requirements.

In the second semester, Jerry applied his accumulated knowledge in his teaching by summarising teaching patterns, such as the selection and application of examples. From his observation of Qian's teaching and implementation of his own, he noticed that teachers could select examples by considering 'the appropriate difficulty level of the examples for students to do' and 'the mistakes that students usually make in doing related exercises'. He summarised three key uses of examples in teaching: (1) to improve students' understanding of new content; (2) to support teachers' demonstration of how to apply new contents; and (3) to provide exercises aimed at helping students grasp problem-solving procedures by imitating the teacher's demonstrations. Jerry particularly emphasised examples in the second and third categories, which were usually selected from the examination question bank; his demonstrations using the examples focused on mistakes that students may be expected to make.

> When students encountered similar exercises in their homework and tests, they recognised that it was important to learn these and listen carefully to the teacher's classroom instruction.

During the third semester, Jerry distinguished between the different roles that content played in the curriculum, particularly with respect to the *Gaokao*, based on his school and instruction experience, his learning of the teaching materials at hand and his communications with other teachers. He then implemented detailed demonstrations of important content in class, based on his understanding and interpretation thereof. Jerry reported that he sometimes adopted Sun's approach to introducing important content, which focused on connecting the content to other contents, because the *Gaokao* typically examined the integrated use of mathematical knowledge. This was the key aspect that Jerry learned from Sun and borrowed for use in his own teaching. Jerry also selected mathematical examples involving the integrated use of knowledge to strengthen students' practice. Although he trusted the textbook's instructional design and used it in his instruction, he thought that the difficulty level of the exercises in the textbook did not meet the *Gaokao*'s requirements.

5.4.4 Summary and Discussion

At the medium stage, Jerry's teaching continued to emphasise the teaching content and focused on students' examination performance. His implemented teaching appears to have been influenced by the environment, emphasising students' acquisition of the content and skills required to complete exercises relevant to the examination.

Interactions between teacher beliefs, knowledge and practice
Compared the teacher beliefs Jerry held at the beginning stage, his teaching at the medium stage obviously reflected more traditional beliefs about teaching and learning mathematics, with emphases on the importance of examination and students' practice to achieve good examination performance. His teaching practice focused on the teaching of important examination-related contents, and students' acquisition of the problem-solving skills needed for the examination by imitating his demonstrations. In the meantime, he had been learning about the knowledge of the contents and students, according to examination requirements, as well as teaching strategies such as using mathematical examples and offering individual tutoring. With the successful implementation of these teaching strategies, it was predicable that Jerry would have more confidence in teaching for the examination, and his examination-oriented beliefs would also be bolstered (Bandura, 1997; Buehl & Beck, 2014). Thus, there was an implicit reciprocal relationship between the three elements.

Environmental influences
The school appears to have directed and supported Jerry in implementing teacher-centred, content-focused and examination-oriented teaching, mainly through one-to-one mentoring and collective teaching activities initiated by the LPG. Jerry learned about the curriculum, examinations, teaching contents, and classroom management mainly by observing Qian's and Sun's classroom teaching and by asking their opinions. This may also have impacted his teaching beliefs; for example, his emphasis on student discipline to strengthen teacher authority

and his particular focus on students' performance can both be considered more traditional beliefs (Stipek, Givvin, Salmon, & MacGyvers, 2001).

During the third semester, Jerry's teaching appeared to be heavily influenced by the environment and its strong emphasis on the *Gaokao* examination. Jerry attributed this to the change in LPG leadership, as Teacher Sun had taught students to prepare for the *Gaokao* for many years. Jerry also reported that it was very difficult for all of the group members to agree on specific issues, because of their diverse teaching and leading experiences, implying that the members already emphasised the importance of the examination. Other researchers have reported that public examinations in China affect teachers' teaching by exposing them to intense pressure from students, parents and other educators (Ma, Lam, & Wong, 2006). Thus, Sun's encouragement and measures may merely have reflected reality. In this situation, Jerry's teaching focus was on his students' performance in the unified examinations. He used the collective lesson plans and unified the student exercises provided by Jerry-LPG2 and emphasised practising exercises that were relevant to the examinations.

5.5   The End Stage of the Two-year Professional Learning

The supplementary data used to triangulate the interpretation of Jerry's professional learning included his reflections on his own learning and his comments on his two mentors, Qian and Sun, both of whom declined to participate in this study. In Sun's case, he felt the investigation was a waste of time, and he already had a heavy teaching load; as the school did not make his participation compulsory, he opted not to be involved.

5.5.1   Main Features of Teaching Practice

At the end stage of the two-year period, Jerry continued to use the collective lesson plans but modified them with his understanding of the contents, obtained by studying the textbook and numerous related exercises. In his teacher-controlled and structured classroom instruction, he also emphasised students' practice and invited individual students to interact with him by asking them factual questions. He reflected on his teaching less often, because he believed that self-reflection was unnecessary for examination-oriented teaching, although it may be important for novice teachers' development.

5.5.1.1 Lesson Preparation

Jerry still used the collective lesson plans during this semester but independently modified the lessons. He did not observe Sun's classroom instruction because, as he stated, the teaching delivered to the students in the Science Class was inappropriate for the students in the parallel class, and he was lacked to energy to invest extensive amounts of time in more observations. The collective lesson plans simply provided him with the contents required to teach and some mathematical examples; he still required a comprehensive understanding of the content and how they should be taught. Thus, he referred to the textbook and reviewed numerous related exercises. He reported that he trusted the instructional design presented in the textbook.

> The textbooks were written and edited by experts and recognised by the Board of Education. They (the instructional designs) must be persuasive and have been testified. So it was better to follow the instruction presented in the textbook, rather than the individual teacher's own design, as it was hard for the teachers to transcend the experts.

Reviewing related contents helped him to identify each lesson's 'core content' and to determine which mathematical examples to use. In Jerry's opinion, the examples should focus on the core contents, and the new contents should be connected to other mathematical topics.

5.5.1.2 Lesson Implementation

Jerry consistently implemented structured and teacher-controlled classroom instruction and introduced new content based on its connections with previously learned content. He also emphasised that students should practise to consolidate their understanding of the contents and their applications. One minor difference was that the majority of questions he posed required factual (51.99%) rather than yes/no answers (see Appendix 9.1.2).

5.5.1.3 Lesson Reflection

Jerry reported that he seldom reflected on his teaching at the end stage because reflections seemed less important given the examination-oriented teaching he was delivering at that time. The teaching was based on rules or patterns he had observed elsewhere, such as how to prepare the lessons and how to choose mathematical examples, and the school did not oblige him to reflect:

> Luckily, our school does not put too much pressure on us to do so, as long as you
> (the teacher) can promote students' performances in examinations. And I had become
> lazier than I was in the first year.

His third explanation was that he could not assess the effectiveness of his
reflections in time.

> Even if I came up with some ideas based on reflection, such as more appropriate
> teaching methods for specific mathematical topics, I could not implement them imme-
> diately. How can I expect to remember it after two years when I come to teach the
> same topic again?

He admitted, however, that reflection was useful for novice teachers, as it helped
them to learn about teaching from various perspectives; 'by reflecting on the
teaching content, [teachers] may understand more about the curriculum, which
may then work in their future teaching'. However, neither good preparation nor
reflection on teaching was important for the promotion of students' examination
achievements: 'wonderful classroom instruction is only part of the teaching, while
the key lies in follow-up practice'. Thus, he preferred to invest more time in
encouraging students to engage in extensive practice.

5.5.2 Jerry's Teacher Beliefs

5.5.2.1 Stated Beliefs

Jerry's stated beliefs distinguished between the mathematics taught in school and
the mathematics involved in research (see Appendix 9.2). First, he believed that
'mathematics consists of various facets' but that of these 'logic and mathematical
thinking are the most essential' and that 'mathematics for doing research in the
subject [of mathematics] should be a process of inquiry'. However, the mathematics taught in schools, Jerry thought, was static, because it constituted 'a set
of rules' to be acquired by students with the guidance of teachers. Thus, Jerry
held mixed beliefs about the nature of mathematics for schooling, although he
considered the subject *per se* to be dynamic.

Jerry emphasised the importance of students' understanding and practice for
the teaching and learning of mathematics as well as the teacher's role in 'guiding
students to learn'. However, his way of 'guiding' was limited to 'telling [the stu-
dents] the way that you (the teacher) learnt mathematics and your understanding

of the knowledge'. Moreover, he relied heavily on practice to consolidate students' understanding and reinforce their memorisation of the contents. He also admitted that the ultimate goal of his emphases on students' understanding and practice was students' good examination performance, which he considered the foremost objective of his teaching. On the other hand, although he recognised the importance of students' interest in learning mathematics, he chose to ignore it because the school did not pay attention to the promotion of students' interest.

> I don't care about whether [the students] learn mathematics well or not. I only focus on their achievement. My mission is to help them get good achievements (in exams). If talking about the interest (students' interest in mathematics), who cares about it in our school?

5.5.2.2 Enacted Beliefs

Jerry's teaching reflected his emphases on content and students' performance. He delivered well-designed curriculum materials (i.e., the textbook) and particularly emphasised the 'core content' that he obtained through reviewing exercises in supplemental teaching materials. According to extant local literature (e.g., Cai, 2012), these exercises are primarily related to topics frequently featured in examinations.

Jerry's classroom teaching also focused on the connections between mathematical topics to promote students' understanding of the content and on the content that students typically struggled with when completing relevant exercises. In addition to his teacher-controlled instruction, all of these features revealed the fixed beliefs he enacted in teaching.

5.5.3   Jerry's Teacher Knowledge

5.5.3.1 'Discussing Mathematical Problems'

Compared with the same tasks Jerry completed at the beginning stage, he continued to focus on the mathematical topics involved in the problems and how he would teach the problems (see Appendix 9.3). He particularly mentioned the mistakes students would make in doing problems and the strategy of asking students questions to draw their attention to their mistakes and to the problem-solving procedure. In addition, he mentioned that he sometimes would consider a problem's 'variation' (变式 *bianshi*), an approach to teaching mathematics investigated by Gu (1991) and advocated by the District Teaching Research Office (TRO). Jerry reported that he learned about 'variation' through teaching activities organised by the district TRO and believed that teaching mathematical problems using variation could help students systematically acquire problem-solving strategies for a set of similar problems. However, he also thought that teaching with variation was not improvisation and required teachers to review and summarise a significant number of exercises in preparation and to complete more exercises with the aim of accumulating knowledge of various types of mathematical problems that were typically examined in the *Gaokao*.

5.5.3.2 Knowledge in Teaching

The content (*combination*) Jerry taught at this stage was considered key content for the *Gaokao*; thus, he was careful to provide clear and detailed demonstrations thereof. For example, when delivering lessons on *permutation* and *combination*, he drew on his knowledge of content, students, the curriculum and teaching to focus on the important and complex distinctions between the two concepts and provided increasingly difficult examples to help students practise their use. He also demonstrated the derivation of related formulae in detail to ensure students' profound understanding or even acquisition of the mathematical procedures.

As Jerry reported in the interviews, he was familiar with the procedure for teaching a mathematics lesson in class. It started with introducing new concepts, including their contents and the derivation process, followed by examples and exercises to consolidate students' understanding of the concepts and to teach them to use the mathematical knowledge to solve mathematical problems, especially those common in examinations. As mentioned above, Jerry relied on supplementary teaching materials to identify the important and difficult points in each lesson, and made use of his knowledge of the students to make his teaching proceed smoothly.

> In the two years' teaching, I know exactly who I could ask to when I have a question
> in the classroom teaching.

5.5.4 Jerry's Self-reflection on His Learning and Comments on the Mentoring

5.5.4.1 Jerry's Self-reflection on His Learning

Jerry's reflection on his learning focused on his pedagogical beliefs and knowledge that were strengthened, acquired or accumulated during the two-year period. The strengthened pedagogical beliefs lay in his emphasis on promoting students' performance in examinations; he believed that his efforts to push students to practise were essential to the effectiveness of his teaching. Rather than delivering wonderful lectures, Jerry preferred to spend more time on individual counselling for students: 'if time allows, I would like to ask every student to my office'. With each lesson taught, he said, he had learned more about the important content and relevant mathematical problems usually examined in regular tests, owing to his acquisition of various kinds of knowledge, including the knowledge of content, curriculum (in particular, assessment) and students. For example, Jerry reported that he had gradually become familiar with his students during his teaching, which had contributed to his ability to judge the appropriate level of content difficulty.

Jerry also stated that the most immediate motivation for his learning of teaching lay in what students should and really could learn in class. Thus, the teacher should learn various kinds of knowledge that they then incorporate into their teaching in each lesson. He also pointed out that the school pressured him to teach with students' academic achievements as the ultimate goal.

> If you did not teach well, the students' achievements were not good. Other teachers would talk about it, and the principal would talk to you. I, myself, cannot accept it either.

5.5.4.2 Jerry's Comments on the Mentoring

Jerry considered mentoring to be a good opportunity for him, as a novice teacher, to learn from the school's more experienced teachers. In Jerry's mind, his two mentors, Qian and Sun, showed him different styles of teaching in the two years; however, their students' academic achievements were not significant different, based on the unified tests. This strengthened Jerry's belief that teachers' lectures are less important for students' examination performance.

Jerry highly valued Qian's profound understanding of the contents, and his ability to manage students. He learned from Qian how to teach specific mathematical topics, and emphasised students' discipline as Qian did. From Sun, Jerry learned about the connections between mathematical topics. As Jerry reported, Sun focused particularly on the connections between curriculum contents, because the *Gaokao* usually examined students' integrated use of mathematical knowledge to solve mathematical problems. Jerry also pointed out that the group-based teaching activities organised by Sun also influenced his teaching by providing him with teaching materials, such as collective lesson plans and unified student exercises. This led him to concentrate on the examination-oriented teaching, but did little to help him learn about other teachers' teaching.

No follow-up learning activities were provided for the collective lesson plans or any other aspects of teaching. Jerry explained that this was because the teachers did not cooperate with one another; most held leadership positions in the school and struggled to to take on board others' suggestions or comments.

5.5.5 Summary and Discussion

At the end stage, although his lessons were based on the collective lesson plans, Jerry prepared them independently by studying the textbook and related exercises. In class, he continued to employ teacher-controlled instruction to deliver the curriculum materials (i.e., the textbook), and relied on the connections between mathematical topics to introduce new contents. He also promoted students' examination achievements by encouraging them to practise, both in and out of class. The positive results derived from encouraging practice, as evidenced by students' examination achievements, led Jerry to believe it was unnecessary to spend much time preparing lessons and reflecting on teaching. These features of Jerry's teaching appeared consistent with his beliefs. Although he believed that mathematics for research should be a process of inquiry, he considered mathematics for schooling to be static. He focused on students' understanding and practice, and believed that their purpose was to promote students' performance. In particular, he also claimed that students' performance was more important to his teaching than their interest in learning mathematics and attributed this attitude to the influence of the school environment.

The section that follows summarises and discusses Jerry's learning outcomes, and his strategies for achieving his learning outcomes across the entire two-year period.

5.5.5.1 The Overall Learning Outcomes

Jerry consistently held mixed beliefs about the nature of mathematics that focused on the static content of the mathematics curriculum, the contents' connections to other mathematics topics and students' understanding of both. He initially emphasised the interactions between mathematics and the real world. Later, however, while he still admitted the dynamic nature of mathematics for doing research, he particularly pointed out that the mathematics taught in schools was static. Jerry consistently emphasised students' understanding and academic achievements in the teaching and learning of mathematics. During the two years, it appears he put increasing emphasis on students' examination performances and the role of practice in students' achievements.

Correspondingly, Jerry's teaching practices increasingly focused on students' examination performance. He believed in the authority of experts, as revealed in his willing obedience of experienced teachers and textbooks. He emphasised the detailed explanation of content, particularly important curriculum content, core lesson content relevant to the examinations, and content that the students struggled to grasp. He also relied extensively on the students' practising for them to master the types of questions usually found in examinations. Over the two years, Jerry consistently implemented teacher-controlled instruction with numerous easily answered questions; however, at the beginning stage, he appeared not to have much knowledge for teaching, and so he imitated his mentor's teaching. Later, he became relatively independent in preparing his lessons.

From the knowledge reflected in Jerry's teaching and his own reflections on his professional learning, it appeared Jerry had acquired the knowledge of teaching, content (for curriculum and examination purposes) and students. This acquisition of knowledge supported the teaching goal that formed his primary focus—the promotion of students' academic achievements.

5.5.5.2 Jerry's Approach to Achieving His Learning Outcomes

The professional learning process that ultimately led to Jerry's examination-oriented teaching was similar to Doris's professional learning model (Figure 4.2 in Chapter 4), which focused on the teaching of school mathematics. Figure 5.1 below illustrates the interactions among the individual elements as well as the influences of Jerry's previous experiences and the environment in which his teaching took place.

The investigation of the beginning stage of Jerry's two-year professional learning indicated that the teaching-related experiences (e.g., teacher training and personal experiences in schooling) that he brought to bear on his teaching influenced his pedagogical beliefs and knowledge and thereby his teaching practice. However, it appeared that these experiences were insufficient to enable Jerry to independently implement his teaching at the beginning of his career, particularly his teaching of specific content. Thus, he relied considerably on the school's support in the form of mentoring, which allowed him to imitate the classroom teaching of and receive one-to-one instruction from experienced teachers.

Combined with the findings on his teaching at the later stages, it appeared the various supports—teaching materials, mentoring, collective teaching activities, etc.—from the school or other organisations led Jerry to gradually believe in and adopt an examination-oriented teaching approach while relinquishing his more non-traditional beliefs (e.g., emphases on students' interest and efforts).

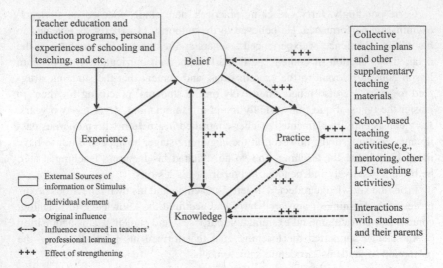

Figure 5.1 Jerry's professional learning focused on the examination-oriented teaching

The entire investigation indicates that consistent interactions between teaching experiences, knowledge, beliefs and practices resulted in Jerry's individual professional learning in three aspects—teacher beliefs, knowledge and teaching practice—and would impact his future teaching. However, the environmental influence in School B seemed to have a greater impact on his teaching practices than the experiences that he brought to bear on his teaching. In particular, it directed him towards examination-oriented teaching through intervention in his daily practice.

The Case of Tommy

6

6.1 Introduction

This chapter describes the professional learning of Tommy (pseudonym), who worked in School C, a key upper secondary school in District III. He obtained a bachelor's degree in mathematics and applied mathematics, and a master's degree in mathematics from a prestigious comprehensive university. As he reported in the interview, Tommy underwent no formal training in teaching prior to working in School C. During the two-year teaching period, he had to focus particularly on learning how to deliver teaching in the school context. He taught two groups of students with significantly different academic abilities over the two years, which led him to recognise the importance of taking the students' individual characteristics into consideration.

6.2 Background

6.2.1 Tommy's Background

Tommy was confident that he had accumulated rich, advanced mathematical knowledge during his bachelor's and master's programmes (over seven years), but did not obtain any knowledge of teaching. Although he had experience in teaching part-time tutorials for secondary students' mathematics over his seven years of study, he did not think his teaching in School C benefitted from it.

Supplementary Information The online version contains supplementary material available at (https://doi.org/10.1007/978-3-658-37236-1_6).

The perceptions, expectations and imaginations about schooling and teaching for mathematics that he brought to his career were drawn from his own experiences as a secondary school student.

After reaching an agreement of employment with School C, Tommy commenced a one-semester internship at the school (from March to June, 2013), during which he observed the teaching conducted by Zhou, a 'senior teacher'[1] who was later one of Tommy's two mentors during his first year of teaching. Tommy corrected the homework completed by Zhou's students and sometimes taught Zhou's classes, but only under Zhou's guidance and only for exercise lessons in which he explained how to solve selected mathematical problems. The internship also allowed him to discuss with Zhou any questions that his teaching raised, to understand how daily teaching practices were implemented in School C and to recognise that an experienced teacher's focus when teaching specific content differed considerably from his own interpretation of the same content.

As Tommy reported in the interview, he became a teacher because he was attracted to teaching as a career based on the fact that teaching has a sense of mission, much like medicine or police work. Tommy further said that he liked the relatively simple and stable working environment of teaching.

6.2.2 School Information

Like School B, School C is an old key upper secondary school, established more than 50 years ago. It consists of two departments: the regular secondary school department and the international department. Tommy was recruited by the regular department, which constitutes Grades 10, 11 and 12, each of which has four class categories: special, parallel, collaborating and Xinjiang classes. Most students in these classes aim to pursue further education in universities or colleges through the *Gaokao*. Students in both the special and parallel classes are independently recruited by School C, while the students in the collaborating[2] and Xinjiang classes[3] are recruited by School C with the cooperation of and to promote education in special regions. Students in special classes are regarded as

---

[1] Senior teacher is a professional title equivalent to Associate Professor at a university (Fan, Miao, & Mok, 2014). District III required that at least one mentor be a senior teacher.

[2] The establishment of collaborating classes followed a contractual program for School C to help improve students' achievements in a certain district in Shanghai.

[3] The establishment of Xinjiang classes aimed to respond to the State Council of China's call from 2000 in Shanghai to promote the talent development of minorities in the province of Xinjiang. http://shmzw.gov.cn/gb/mzw/shmz/mzjy/userobject1ai447.html.

having the best academic abilities among their peers in the same grade. They are first assessed by School C through examinations and interviews before the *Zhongkao* (autonomous enrolment) but are not admitted if they fail to meet the school's minimum *Zhongkao* achievement requirements. Students in Xinjiang classes study the same mathematics curriculum as local Shanghai students and, like them, will also eventually be assessed by the *Gaokao*. However, they have a longer time period (four years) in which to complete their upper secondary school learning, while local students must complete theirs in three. Moreover, universities have limited Xinjiang admission quotas, and Xinjiang students compete for university admission only with other Xinjiang students in Shanghai (rather than all students).

Tommy taught one parallel class and one collaborating class in Grade 10 during the first semester but only taught the parallel class in the second. During the second year, he was assigned to teach one Xinjiang class in Grade 10. The analysis of his teaching in this chapter focuses on the data from his parallel class (Tommy-Class1) and the Xinjiang class (Tommy-Class2). Tommy did have additional teaching duties during the two years, including teaching an optional mathematical competition class in both years and one class in the international department (as a substitute teacher) in the second year. He thought the heavy teaching load made him too busy to reflect on his teaching frequently.

Tommy was involved into two Lesson Preparation Groups in the two years, called Tommy-LPG1 and Tommy-LPG2, respectively. All group members shared the same office, which enabled them to freely discuss daily teaching with others, rather than rely on overly formal collective teaching activities. The LPGs also arranged the unified teaching schedule, students' exercises, and progress examinations to coordinate the same teaching pace for each member. As Tommy reported in the interviews, his teaching was mainly influenced by his interactions with the group leaders through their mentoring. Li, the leader of Tommy-LPG1, and Zhou were assigned as his mentors in the first year. As Zhou did not teach in the same grade as Tommy, he fulfilled his mentor obligations by regularly observing Tommy's classroom teaching and providing feedback and suggestions thereon. It was required by the district induction program that Zhou should observe Tommy's classroom teaching once a week. However, according to Tommy, 'he (Zhou) usually forgot to come, because he was too busy with teaching for Grade 12 students who would sit the *Gaokao* soon'. Tommy had more opportunities to communicate with Li, and was attracted to Li's lively and interesting classroom teaching as well as his comprehensive mathematical knowledge. In learning from Li, he usually focused his discussions on how to solve difficult mathematical problems or introduce specific contents. Even during the second year, he continued to attend

Li's class to observe how Li taught new content that he had not yet taught but would teach in the future, because he was still teaching Grade 10 students and Li was teaching Grade 11. At that time, Wu, the leader of Tommy-LPG2, was his mentor; Tommy said that the main thing he learned from Wu was how to teach Xinjiang students.

6.3   The Initial Stage of the Two-year Professional Learning

6.3.1   Main Features of Teaching Practice

During the early stages of Tommy's career, he relied on a combination of his own knowledge of the content (from his previous learning experiences) and Li's interpretations (in the real context of teaching) when preparing his classes. In the classroom, he implemented teacher-controlled instruction, which conveyed his interpretations of the content to students using various examples. Although he endeavoured to deliver his teaching in alignment with the unified teaching schedule, he did not pay attention to the capacity[4] of each lesson. In reflecting on his teaching, he felt that it was dull and attributed this difference to the influence that he had far less teaching experience than Li.

6.3.1.1 Lesson Preparation

When preparing his lessons, Tommy first wrote up lesson plans independently, based on his own interpretations of the content and with reference to the teacher's book and other lesson plans that he found online. He stated that he had confidence in his interpretation of the content, owing to his own outstanding learning achievements in the same content as a secondary school student and his extensive mathematics learning. He particularly emphasised the logical relationships between various elements of mathematical content, which he considered essential for the construction of teaching procedure. However, after observing Li's teaching, he found that his subject knowledge alone was insufficient owing to his lack of pedagogical knowledge. As Tommy reported in the interviews, prior to teaching, he had 'not recognised that teaching specific content was so different from learning that same content' nor that a teacher had to consider 'not only the relationships between contents', but also 'which points in the contents students

---

[4] Lesson capacity (课堂容量) refers to how many topics (or how much knowledge) should be taught in a single lesson (same as footnote 3 in Chapter 4).

may find difficult to understand'. Tommy also claimed that it was impossible for him to acquire all the necessary knowledge immediately. He was eager to experience teaching, because it would allow him to acquire the necessary skills and knowledge. At that moment, he decided to convey all his interpretations of the contents to the students in the classroom.

6.3.1.2 Lesson Implementation
Tommy explained content in detail, using various examples to convey his interpretations to the students. He also paid attention to the pace of his teaching, endeavouring to follow the unified teaching schedule, but did not control the capacity of each lesson (Appendix 10.1.1) as did the subjects in the other two cases. He usually ceased his instruction when the class time ended, regardless of the integrity of his explanation of a certain concept. His classroom instruction was teacher-controlled and included numerous easily answered questions (60.81% yes/no questions and 23.83% factual questions) that Tommy posed to the entire class to lead his instruction (Appendix 10.1.2).

6.3.1.3 Lesson Reflection
Based on his reflections on his own classroom teaching and his observations of Li's instruction, Tommy believed that his teaching was too dull, since he focused excessively on the content without providing any highlights. He observed that he had no choice but to teach in this way because he knew too little about the students, including what they had already learned and what level of difficulty they could understand. Under the circumstances, although he hoped to mobilise students to think in their mathematics learning, he did not know how to achieve this in practice. However, Tommy remained hopeful, believing that these problems could be solved through the accumulation of teaching experience. Thus he declared that he would do his best to learn how to teach by practicing his teaching and engaging in interactions with Li.

6.3.2 Tommy's Teacher Beliefs

6.3.2.1 Stated Beliefs
Tommy appeared to hold non-traditional beliefs about the nature of mathematics. He acknowledged mathematics' dynamic nature and regarded it as a process (see Appendix 10.2). Tommy conceptualises mathematics as a process that originated with a few axioms and then was developed by human beings through

rigorous logic. Thus, mathematics consists not only of axioms but also of 'logical reasoning and products of the reasoning'. Tommy's teaching and learning beliefs consistently focused on learners. Based on his own experiences in learning mathematics, he valued the roles that students' interest, talent, thinking and understanding of mathematics play in mathematics learning and believed that teachers should make efforts to promote these.

Tommy also asserted that mathematics teachers should have outstanding mathematical knowledge before they commence teaching, as without it they may misunderstand the mathematical content. In addition, he believed that pedagogical knowledge is relatively less important as it can be learned naturally during the course of teaching.

6.3.2.2 Enacted Beliefs

However, the beliefs that Tommy enacted in his teaching were more traditional. He focused on the contents, emphasised students' understanding, and his main teaching strategy was to explain the content in the greatest detail. He stated in the post-class interviews that he used the connections between the mathematical topics to organise his classroom teaching and carefully illustrated all facets of each concept in class. By 'facets', Tommy meant the content points that he emphasised in his teaching, which generally included key aspects of a given mathematical topic that students must know or in relation to which they frequently make mistakes when completing exercises.

Tommy also conveyed his non-traditional beliefs about mathematics to his students by introducing Russell's paradox when teaching 'sets'. He told his students that mathematics is systematic and constructed and encouraged them to learn some mathematical concepts by reading the textbook themselves. However, Tommy only permitted autonomous student learning for some simple mathematical concepts; his main teaching practice remained teacher-centred and content-focused.

6.3.3 Tommy's Teacher Knowledge

6.3.3.1 'Discussing Mathematical Problems'

In discussing mathematical problems (see Appendix 10.3), Tommy focused on the mathematical topics involved in the problems and then turned his attention to students by, for example, pointing out errors that students commonly make regarding mathematical topics when doing related exercises. Based on his own learning experiences, he also proposed that students should acquire a solid mathematical knowledge base and develop problem-solving strategies (e.g., simplifying or transforming problems into familiar forms). He particularly mentioned certain aspects of mathematical thinking (i.e., the method that combines algebra and geometry) that could be used when solving problems. Tommy appears to have focused mainly on subject content knowledge, including the content in the problems, and how to solve the problems.

6.3.3.2 Knowledge in Teaching

As mentioned above, Tommy used the connections between mathematical contents to construct his teaching and conveyed his interpretations in class using various examples. Thus, in implementing his teaching for each lesson, he showed the content knowledge he obtained from his own learning experience or from his observation of Li's classroom instruction, and integrated his specific mathematical knowledge into his teaching. For example, when teaching sets, he introduced the origin of the abbreviations of the common number sets (e.g. the real number set) to help students memorise these number sets easily.

Tommy also showed his knowledge of pedagogical content and mathematical content by using various examples in his classroom teaching. For instance, the teaching strategy of metaphor (e.g. comparing sets to the files in the computer) can be seen in his illustration of some mathematical concepts with examples from the daily life. To illustrate that 'sets themselves can be members of other sets', Tommy even introduced the example of Russell's paradox, which is not required in elementary mathematics. In addition to Russell's paradox, he used two other examples. The first example, a set consisting of a pile of books and a box of chalk, was used to convey the general idea to students. When Tommy used the second example (a set including the class and Tommy himself) to determine whether the students had understood the point, they incorrectly responded that they were the elements of the sets, since their common sense told them they all belonged to the set member 'the class'. Tommy simply repeated the first example to explain the point several times, indicating Tommy's ignorance of pedagogical techniques other than explanation.

Although Tommy conveyed mathematics' dynamic nature to the students using Russell's paradox as an example, he did not do so purposively, according to the post-class interview. Actually, he recognised two weaknesses in his teaching, including his use of descriptive explanations of the contents without highlights, and his having no specific pedagogical way to mobilise students' thinking. He reported having no idea of how to solve the two issues in the moment and that he would rely on the accumulation of teaching experience to do so.

6.3.4   Summary and Discussion

Initially, Tommy organised his teaching based on his understanding of the mathematical contents, including his interpretations of each content item *per se* and the connections between mathematical topics. He then employed descriptive content explanations to realise students' understanding in class. The classroom teaching that Tommy implemented was teacher-controlled and in line with the unified teaching schedule. However, his enacted teaching beliefs appeared consistent with the mixed category, since he also emphasised mathematics as a process of inquiry to the students in classroom teaching. Tommy's teaching indicated that he had knowledge of the mathematical contents but lacked pedagogical knowledge. As his reflections on his teaching indicate, he recognised his weaknesses and attributed them to his lack of teaching experience.

Tommy's stated beliefs were quite different from his enacted beliefs. In his stated beliefs, he considered mathematics to be a process of inquiry and emphasised the importance of students' interest, thinking and understanding to teaching and learning in mathematics, an attitude that approaches a non-traditional stance. In particular, Tommy emphasised that the teacher's mathematical knowledge was more important to mathematics teaching than pedagogical knowledge. In discussing mathematical problems, he focused on the mathematical knowledge involved that students required for solving problems, including relevant mathematical facts, mathematical thinking and problem-solving strategies.

6.3.4.1 Individual Level Orientation to Learning

The role of experience
Both learning and teaching experiences appeared to influence Tommy's teaching. Based on his own elementary and advanced mathematics learning experiences, Tommy emphasised the dynamic nature of mathematics, learners' efforts, and the importance of teachers' mathematical knowledge in his stated beliefs. He also relied on interpretations of the contents he obtained from his own learning, indicating that his learning experience contributed to his original beliefs and knowledge for teaching.

However, Tommy also attributed his preference for teacher-centred teaching and his simple approach to transferring content knowledge to his lack of teaching experience. As he believed that pedagogical knowledge can be acquired through the accumulation of teaching experience, he looked forward to learning to teach while teaching. He appeared to believe that novice teachers require teaching-related experience early in their careers to be able to teach.

The influences of teacher beliefs and knowledge on teaching practice

Teacher beliefs have been widely recognised as indicators or even determinants of teachers' teaching practice (Bandura 1986; Dewey 1933; Pajares, 1992; Thompson, 1992). However, past studies (e.g., Raymond, 1997) have observed inconsistencies between mathematics teachers' beliefs and practices, and researchers have suggested several reasons for these inconsistencies, ranging from the individual teacher to the situated context (Ernest, 1989; Thompson, 1992; Beswick, 2007). In Tommy's case, significant contradictions were evident between his stated and enacted beliefs. He reported that he regarded mathematics as a dynamic process and that he believed in learner-centred mathematics teaching and learning but implemented teacher-centred teaching with an emphasis on contents. Referring to Thompson (1992), these contradictions were likely due to Tommy's lack of necessary teaching skills and knowledge.

Tommy's teaching revealed that he lacked pedagogical knowledge, such as the need to have clear purposes in teaching certain content or using a given example, the various approaches to conveying subject knowledge to students and ways to achieve students' active engagement; however, he did appear to have a comprehensive understanding of all facets of the content. Thus, he sought to foster student understanding by explaining his interpretations to the students in as much detail as possible. He transferred as much of his content knowledge to them as possible—not only textbook knowledge but also such additional knowledge as the origins of number set abbreviations, some historical knowledge, and advanced mathematical knowledge (i.e., Russell's paradox). Comparing his classroom teaching to that of his mentor, Li, Tommy recognised that he offered too few highlights and lacked effective ways of mobilising students' thinking. He relied on his ongoing acquisition of teaching experience to improve the situation.

To summarise, Tommy's non-traditional beliefs and comprehensive understanding of content knowledge was influenced by his prior learning experiences, while his lack of pedagogical knowledge, stemming from his lack of teaching-related experience, likely contributed to his failure to implement those beliefs. Like Doris, the inconsistency between Tommy's beliefs and knowledge led us to expect that new learning would occur during the latter stages of the two-year investigation.

6.3.4.2 School/organisational Level Orientation to Learning

As in Jerry's case, School C also began to intervene before Tommy began to practise his teaching by providing a one-semester internship. It also influenced Tommy's teaching through the regulating effect of the unified teaching schedule and students' exercises provided by his LPG. In particular, his mentorship also impacted Tommy's teaching, by affording him opportunities to learn from experienced teachers. Through classroom observations, Tommy could absorb Li's interpretations of the content from a teaching perspective, which he could then implement in his own teaching. He said he learned from his students' feedback and he also recognised by himself that his teaching was dull; comparatively, Li's teaching was more interesting, since there were highlights in a lesson and Li usually successfully promoted students' thinking in doing mathematical problems. In the post-class interviews, Tommy demonstrated his appreciation of Li's rich teaching experience and his desire to improve his own teaching by learning from Li. Tommy appears to have considered Li as his role model. His interactions with Li enabled him to learn more about teaching while simultaneously influencing his daily teaching at the beginning stage.

6.4   The Medium Stages of the Two-year Professional Learning

6.4.1   Main Features of Teaching Practice

During the second semester, Tommy continued to rely on his interpretations of the contents to organise his teaching but with the use of Zhou's lesson plans. In class, he implemented teacher-controlled instruction and focused on the introduction of new content. He felt more comfortable with teaching and believed that he had learned the teaching pattern of introducing new contents in each lesson. He endeavoured to foster students' profound understanding of mathematical knowledge through clear explanations of the content, rather than focusing on students' good examination performances. His mentors focused on his normative teaching behaviours and reasonable instructional design that was aligned with the curriculum.

In the third semester, Tommy used the lesson plans and teaching experience that he had accumulated during his first year to teach a Grade 10 Xinjiang class. His classroom instruction was still teacher-controlled but followed a slower teaching pace due to the students' weaker knowledge base. He frequently asked students whether they understood his explanations and interrupted exercises to review previous knowledge and students' mistakes. In his post-class reflections, he stated that he could not teach abstract content and had to teach as simply as possible to ensure the students understood. He also generally reflected that his greatest challenge, as a novice teacher, was choosing an acceptable content difficulty level for his students.

6.4.1.1 Lesson Preparation

In the second year, Tommy happily reported that he received his lesson plans for all Grade 10 teaching topics from Zhou. This saved him time in searching for supplementary reference materials online. In preparing the lessons, he usually considered how the mathematical concepts could be incorporated into the lesson plan, including the concepts' meanings and applications. He occasionally made slight adjustments to the lesson plans, such as excluding difficult examples developed by Zhou for teaching special classes; otherwise, he implemented the plans directly in the classroom.

Unlike the other two cases, Tommy continued to teach Grade 10 while in his second teaching year but only to students in one Xinjiang class, whose members were, as mentioned above, recruited from Xinjiang Province to promote minority education. The mathematical content taught to the Xinjiang students was identical to that taught to other students in Shanghai, but the teaching was delivered at a slower pace. This allowed Tommy to reuse his existing lesson plans, and he believed that he had improved his understanding of the content and that this had benefitted his teaching since he first used the plans. He said that his observations of Li's classroom instruction contributed to his understanding of such concepts as viewing content interpretations from the students' perspective (e.g., which content might be difficult for students to understand).

6.4.1.2 Lesson Implementation

During the second semester, the LPG's whole-semester teaching plan was to teach the content more quickly at the beginning and then to concentrate on students' errors when they tackled problems later. Thus, in the observed lessons, Tommy mainly focused on the introduction of new content. As he stated, he followed the teaching procedure that he had summarised from his teaching practice—verbally describing the concept, illustrating it, and then consolidating students' understanding through exercises. He still used direct explanations to demonstrate the mathematical concepts to the students and held the authority to initiate questions and determine correctness, but the vast majority of questions remained unanswered by the students. His classroom instruction was still teacher-controlled.

In the third semester, Tommy taught the same content to the Xinjiang class that he had taught to his first-year class, albeit at a slower pace, because he had more class time (two more class periods than other classes). In class, he continued to implement teacher-controlled instruction and explained content to the students in detail. Most of the questions he asked were yes/no questions designed to lead his instruction (52.73%), regardless of student response (see Appendix 10.1.2), however, it was found that the proportion of questions answered by the entire class increased (from 27.66% to 31.61%). Moreover, Tommy asked students multiple times during each lesson whether they understood his illustrations, indicating that he lacked confidence in his instruction. He also invested extensive class time in reviewing existing knowledge and students' errors to strengthen their memory of the content already taught.

6.4.1.3 Lesson Reflection

In the second semester, Tommy believed that his most significant improvement lay in his feelings about his teaching. He was not nervous and felt more comfortable with his daily teaching. He also recognised that people surrounding him were paying a lot of attention to students' examination performance; however, rather than focusing on the skills needed to address examination questions, he continued to emphasise the promotion of students' understanding of the content. For example, he usually focused on clear explanations of how mathematical concepts emerge and are applied, despite acknowledging that these would not directly lead to students' good achievements.

Tommy stated that he mostly reflected on his mentors' critical comments about his teaching, which focused on his normative classroom teaching behaviours and the structure of his instruction. His mentors pointed out that Tommy's blackboard presentations were poor and that he spoke with an accent or a lisp. Tommy thought that, because he had received no formal teacher training, he was obliged to focus on improving his teaching behaviours. The structural problems in his instruction lay in his sequence of teaching and his focus on teaching content, which required him to have a comprehensive understanding of the curriculum. Tommy reflected that he may have been too subjective, as he depended too much on his own interpretations of the contents. For example, he was reluctant to teach about directed segments in the topic trigonometric ratios, despite it being required by the curriculum, because the topic had confused him in his own learning experience. As such, he did not prepare the lesson well in advance of class and thus ultimately taught it incoherently, as he admitted in the post-class interview.

In the third semester, Tommy found that teaching the Xinjiang class was different, because the students had a weaker knowledge base than other students, meaning that he had to teach the content more slowly and with greater patience. On reflection, he thought that the only purpose of his teaching at that time was to teach students how to apply knowledge and to 'help them acquire the procedure of using the knowledge to solve the problems'. As he reported, 'it was better if the teaching could be simpler' and he usually 'excluded the teaching of abstract concepts', since it was difficult for the students to understand his explanations.

Tommy believed that his teaching was usually focused on mathematical content, particularly his comprehensive interpretations of that content. However, it was difficult for him, as a novice teacher, to determine to what extent his illustrations were accepted by his students. He usually found that his students became distracted or even fell asleep when he explained content in detail, which led him to recognise that his teaching could be more effective if he asked his students to practise more or had them demonstrate how to solve problems rather than relying on descriptive explanations. Thus, he planned to learn from his mentor how to illustrate knowledge by emphasising the mistakes that students may make when completing exercises.

6.4.2 Beliefs Reflected in Teaching (Enacted Beliefs)

At the medium stage, Tommy appeared to still hold mixed beliefs and to emphasise content and his students' understanding thereof. In the second semester, he emphasised the development, connections and applications of the content, while in the third semester, he placed greater emphasis on students' understanding and performance.

In the second semester, Tommy still used the teaching strategy of explaining content in detail and focused on the development of, connections between and applications of different concepts in his demonstrations. For example, when teaching the concept of angles, in Tommy-R2-CO1, he illustrated the different ways in which it had been defined in the various stages of their learning—primary, junior secondary and upper secondary school—to demonstrate the concept's development and range of application. Although he found that descriptive explanations were ineffective with respect to student outcomes, he persisted in using them, as he claimed that his teaching aimed at students' understanding of the content rather than their performance.

During the third semester, Tommy continued to focus on content, which he conveyed via teacher-centred teaching. In particular, he placed greater emphasis on students' understanding of the content and their performance than before. In the post-class interviews, he expressed that students should understand rather than memorise mathematical content. He was against some teachers' method of teaching students scripts[5] to memorise mathematical rules, as he believed it would diminish students' interest in learning mathematics. Thus, in his class, he never talked about scripts but instead explained contents in detail to students. He also drew on students' performances in their daily homework and regular examinations to address their misunderstandings. For the Xinjiang class, he had to teach in a simple and straightforward way that directly conveyed the procedure, owing to their relatively weak knowledge base.

6.4.3   Knowledge in Teaching

Tommy's teaching reflected that he had a comprehensive understanding of several topics (e.g., the definition of angle), as well as different levels of knowledge about mathematical curricula, including their concepts and applications; for other topics (e.g. directed segments in teaching trigonometric ratios), his lack of pedagogical knowledge was clear. During the third semester, he taught the same content as in the first semester but to different learners. For effective teaching, Tommy had to learn more about teaching students with weaker knowledge bases.

In the second semester, Tommy's classroom teaching reflected that he had a clear understanding of subject knowledge and sufficient knowledge of different learning stages of mathematics curricula to introduce concepts in a developmental process; for instance, he was able to explain the different definitions of angle in the primary, junior secondary and upper secondary school curricula to help students understand how the concept had developed. At the same time, he used mathematical examples to help explain the content—in particular, its applications. While using the examples, Tommy seldom gave students much time to complete exercises and instead focused on possible mistakes.

Tommy's reflections on his teaching suggest that he relied too much on his own interpretations of the content, based on his own learning experiences, to integrate his content knowledge and the curriculum; however, his pedagogical knowledge remained obscure. For instance, when teaching the content directed segments in the trigonometric ratios topic, he did not clearly understand why the topic should be taught, as his own learning experience told him it was useless. To adhere to the curriculum, he had to teach the content; however, in his classroom instruction, he expressed his own opinion of the content's uselessness. After class, Li criticised him for violating the rules of teaching, saying 'it is not good

---

[5] Such scripts for students to memorise mathematical facts easily and quickly also occurred in Jerry's case and illustrated in footnote 2 in Chapter 5.

for students' motivation to learn mathematics if the teacher himself thinks it is useless'.

During the third semester, Tommy taught the same content for the second time but to students in the Xinjiang class, whose responses and exercise performance betrayed their weaker knowledge base and lower academic ability than other students in School C. He thus adjusted his teaching style, making it slower, simpler and more straightforward. He frequently asked students whether they had understood his explanation and closely reviewed their homework, suggesting that he lacked confidence in his teaching's effectiveness and was still trying to find an appropriate way to teach such a special group of students.

As Tommy reported in the post-class interviews, his observations of Wu's teaching in another Xinjiang class led him to recognise that his comprehensive explanations of the contents were not useful and that he did not need to teach some of the contents to the students at that time. He henceforth considered 'teaching the contents at the right time' to a principle that he must acquire during his teaching.

6.4.4 Summary and Discussion

At the medium stage, Tommy still emphasised the transfer of content and the teacher's role while teaching the two different student groups. To teach the parallel class during the second semester, he emphasised students' understanding of the contents through detailed explanations. To teach the Xinjiang class during the third semester, he implemented the lesson plans that he developed based on his first year of teaching, but in a slower, simpler and more straightforward manner, taking the students' weaker knowledge base and lower academic abilities into consideration.

6.4.4.1 Individual Level Orientation to Learning

Tommy's enacted teaching beliefs appeared consistent with the mixed beliefs revealed in his second semester. Through his teacher-centred practice, he emphasised students' understanding of the content required by the curriculum. While teaching the class, he appeared to draw on traditional beliefs to cater to students with a weaker knowledge base and lower academic ability, emphasising the acquisition of procedures for solving mathematical problems. His interviews suggest that this adjustment stemmed from his students' negative responses to his detailed explanations and his observations of his mentor Wu's teaching. This implies that this teaching practice may have influenced Tommy's tendency to espouse traditional beliefs.

In Tommy's second semester of teaching, he continued to rely considerably on the content knowledge he had accumulated during his prior learning experiences. His teaching in the third semester, however, indicated that his first-year teaching practice contributed to his knowledge, as evident, for example, in his ability to interpret content from a student perspective. Tommy also seemed to acquire knowledge of his students and of teaching through his interaction with students who had a weaker knowledge base and lower academic ability. This indicates a reciprocal relationship between his teacher knowledge and his teaching practice; that is, in addition to his previous learning experiences, his previous teaching experiences also impacted his acquisition and accumulation of knowledge. However, his tendency to select simpler, more straightforward methods of demonstrating mathematical procedures suggests that he still lacked knowledge of various other teaching approaches.

6.4.4.2 School/organisational Level Orientation to Learning

As above-mentioned interactions among Tommy's experiences, beliefs, knowledge and teaching practice indicate, the school environment played an important role in his teaching, especially in the form of mentoring. He seemed to learn how to interpret certain content from his observations of Li's teaching and how to conduct his Xinjiang class by observing Wu's teaching. That is, through mentoring, the school was able to shape Tommy's pedagogical knowledge and thereby influence his teaching practice and beliefs. Moreover, largely due to his mentors' guidance, Tommy adopted the school's normative teaching behaviours, including extensive use of the blackboard, how he spoke in class, and how he designed and implemented lessons.

6.5   The End Stage of the Two-year Professional Learning

6.5.1   Main Features of Tommy's Teaching Practice

At the end stage, Tommy continued to use the lesson plans developed during his first year of teaching but simplified his use of examples. In class, he still implemented teacher-controlled instruction but initiated more interactions with the whole class by posing easy leading questions. He also focused on those topics in which students often made mistakes when completing exercises and paid attention to the procedures needed to solve mathematical problems in his demonstrations. After class, he emphasised the influence of his mentor's instruction and his teaching goal of promoting students' performance in the *Gaokao*, although he had little time to reflect on his teaching *per se*.

6.5.1.1 Lesson Preparation

As Tommy reported in the interviews, he continued to use the same modified versions of Zhou's lesson plans that he had used in his first year. The modifications mainly reflected his study of the textbook and of the students' exercise book (The Practice Book: *yi-ke-yi-lian* 一课一练), which focused on the content and mathematical questions inside. In particular, he also emphasised teaching the content and questions that he had learned from his observations of Li's classroom teaching. While teaching the Xinjiang class, he was obliged to make several adjustments, such as excluding some mathematical examples that he thought were too difficult and simplifying the use of other examples. For instance, some examples were used in the Xinjiang class simply to demonstrate the procedure for solving a certain type of question; however, that same example could also be used to develop concepts when teaching the parallel class. Tommy occasionally attended Wu's class so that he could refer to her way of teaching the content.

6.5.1.2 Lesson Implementation

As in the third semester, Tommy still spent a considerable amount of class time on reviewing previous knowledge related to the content being taught and on demonstrating the procedures for solving questions related to the content. In his demonstrations, he typically emphasised any points that students may have been likely to misunderstand while completing related exercises; he called these 'error-prone points'. To deal with these points, he invited students to participate in his demonstrations by asking numerous easy leading questions (e.g., 49.21% and 35.90% of the questions he asked were yes/no and factual questions, respectively; see Appendix 10.1.2). Students were typically given a related mathematical example to practise so that they could discover the 'error-prone points'. In the classroom, Tommy retained the authority to pose questions and determine the correctness of the students' answers; that is, he continued to implement teacher-controlled instruction.

6.5.1.3 Lesson Reflection

Tommy reported that he had little time to reflect on his teaching owing to his heavy workload. In addition to teaching the Xinjiang class, he had to teach mathematics for an international class because of a teacher shortage, and had simultaneously taken over the running of an optional competition mathematics course. The Xinjiang class also involved weekend classes, during which he would consolidate students' content learning by reviewing their exercises.

In his reflections on his teaching, Tommy reported that he needed to invest effort in improving students' understanding of the content and their acquisition of the skills needed to solve the types of mathematical problems commonly examined by the *Gaokao* because his explicit teaching goal was to help students gain admission to universities or colleges. He focused on the improvements he could make in his teaching, based on his mentors' comments and examples, such as when Wu pointed out that Tommy should reduce his descriptive explanations of content and encourage students to talk.

6.5.2   Tommy's Teacher Beliefs

6.5.2.1   Stated Beliefs

Tommy believed that 'mathematics is the knowledge that [people] must acquire to know the essence of the world' and that mathematics 'contains human intelligence, the basic intelligence' and 'all of the relationships in the world'. He categorised that knowledge into explicit and implicit perspectives. By the explicit perspective, he meant the 'applications of mathematics' in a narrow sense; people usually think that 'mathematics is useless', because 'most of the topics they learned in class were not actually used in real life'. The implicit perspective he considered to be a kind of 'thinking habit' obtained during mathematics learning that people usually ignore (see Appendix 10.2).

Tommy considered the long-term goal of teaching and learning mathematics to be to train an individual's thinking as necessary to their ongoing development. He emphasised the importance of students' interest and thinking, and that 'the most convenient way' to teach mathematics well is to promote students' own interests and encourage them to think while they are learning. He particularly emphasised students' own learning efforts and that teachers' main work should be to pose appropriate problems in due course so as to mobilise students to think. In his opinion, '[Chinese teachers are] used to imparting knowledge to students'.

However, Tommy also stated that the most practical objective for his mathematics teaching and learning in school was to help students enter university, which led to his belief that the goal of teaching was to consolidate students' knowledge base through practices and drill so that they were equipped to pass the *Gaokao*. Given teachers' solid knowledge base, students should follow their guidance to learn the content and complete exercises; at the same time, teachers had to 'supervise students in reviewing and consolidating their knowledge' by practising often and help them to correct the mistakes they make in their exercises.

The 'implicit' perspective of mathematics favours long-term mathematics teaching and learning and implies the closed non-traditional beliefs that Tommy held. The 'explicit' perspective, which is based in practicality, appears to inform the more traditional beliefs that Tommy held in relation to his teaching practice, and in his stated beliefs, when he recognised that the most common teaching method used by Chinese mathematics teachers was imparting knowledge to students.

6.5.2.2 Enacted Beliefs

Tommy's teaching reflected his tendency to espouse more traditional beliefs. He focused on students' acquisition of strategies for solving the problems commonly found on examinations so that they could achieve high marks. In class, he usually summarised the question types for students, provided scaffolding to help them grasp the procedures for solving the various question types and then repeated the solution procedures. He also emphasised the need for students to question each aspect in which they were prone to mistakes. Tommy reported that he frequently helped students to review the errors they made in their daily homework, believing that when students experienced problem-solving procedures dependently, it would consolidate their learning, particularly their memorising of commonly mistaken points. He stated that all of these efforts were aimed at promoting students' achievement of good examination results.

6.5.3   Teacher Knowledge

6.5.3.1 'Discussing Mathematical Problems'

When discussing mathematical problems, Tommy maintained a focus on the mathematical topics involved in the problems but mentioned the examination perspective more often than he had in the beginning stage (Appendix 10.3). His first criterion for reviewing topics was whether they usually occurred in examinations; for instance, the first sub-problem in a second-tier problem was not key *Gaokao* content, in his opinion. He then discussed how to solve the problems and to which content points students should pay particular attention. For example, he said that he had to emphasise the script '逢子必空' (when there is a subset, remember there is a solution of null set) when teaching first-tier problems. He even used scripts to help students avoid commonly made mistakes when doing problems, which conflicted with his medium-stage statement that he was against the use of scripts in teaching.

6.5.3.2 Knowledge in Teaching

When teaching question types, Tommy demonstrated his knowledge of the content, the students and the curriculum, particularly with respect to examinations. He knew what commonly mistaken points were involved in the contents he taught. In the post-class interviews, he said that during his first year of teaching practice, he found that his students had very low academic ability and a weak knowledge base. Thus, in his classroom teaching, he usually illustrated each individual step of the solution procedure for every mathematical problem. In addition, he asked students to perform simple arithmetic or follow simple patterns as a means of involving them in his teaching; for example, regarding the last step of calculating the sum of a series, Tommy found his students had difficulty in telling the value of $3^n \div 3^{n-1}$, which was a mathematical skill taught in junior middle school. He then asked the students the value of $3^9 \div 3^8$; when he had been given the correct answer, he pointed out that $n-1$ was one less than n and then guided students to answer that the value of $3^n \div 3^{n-1}$ was 3.

Tommy found that his students usually struggled to recall the existing knowledge that was expected to be ingrained in their minds. However, he usually could not stop to help students fill in their lack of elementary and junior middle school knowledge, as he had to teach the required upper secondary school content in full. Therefore, he turned to teaching problem-solving procedures in detail and making students memorise them through constant practice, which he considered to be his only course of action. However, he still attempted to promote students' understanding (or at least memorisation) of content by highlighting the connections between mathematical topics. For instance, when teaching the sum of series, he connected the methods for calculating the sum with the derivation of the formulas for the sum of arithmetic and geometric progressions.

6.5.4   Tommy's Self-reflection on His Learning and Words from Li

6.5.4.1 Tommy's Self-reflection on His Learning

Tommy reported that the biggest improvement in his teaching was that it became more normative. He thought his early stage teaching was not wholly professional. The first example he provided was that he thought his blackboard writing was arbitrary. He said he usually wrote out the information he wanted to convey to his students anywhere on the board and even sometimes made a draft of mathematical operations and derivations on the blackboard during class. These usually made his blackboard presentations quite messy and incoherent. He mentioned the importance of clear blackboard demonstration; 'if some of the students cannot follow my instruction, this would allow them to record the procedure and then think it over after class'.

In the second example, Tommy discussed how he selected the content that he taught to students in each lesson. He said that initially, he usually selected content and arranged its sequence by considering the logical connections between mathematical topics, regardless of whether those topics were appropriate for the students to learn or were not required by the curriculum. Tommy found that the inclusion of such additional content usually confused students and that their learning thereof could not be consolidated owing to a lack of provided exercises. Generally, Tommy thought that he did not know much about his students, including where they might have problems and which methods were most effective for their learning. The only references he could use were students' daily homework performance and how his mentors taught the same content in their own classes.

When Tommy's mentors observed his classroom teaching, their comments focused on his normative teaching behaviour—that is, his blackboard presentations, voice and pronunciation, lesson structure and classroom interactions with students. Tommy said that Zhou was more careful; his comments included the exact words he wrote on the blackboard as well as the distribution and arrangement of the contents. Li and Zhou focused on what content should be taught to students and in what sequence during the first year. Wu suggested that Tommy should lecture less and allow his students to talk more in class. At the end of the interview, Tommy said that he would make future efforts to learn normative teaching and his students' natures.

6.5.4.2 Words from Teacher Li (Tommy's Mentor)
The interview conducted with Li examined his stated beliefs as a mathematics teacher, his instruction of Tommy as a mentor and his comments on Tommy's teaching.

Li regarded mathematics as revealing the essential relationships in nature. Li differentiated between his actual teaching of mathematics and his ideal teaching of mathematics. He believed that the best way to learn mathematics was for learners to study independently. The student's talent determines the upper bound for their mathematics learning while individuals' efforts ultimately determine how much they obtain from their learning. He also claimed that ideal teaching should not be delivered on a national scale, but only with students who are able to learn independently with teacher supervision and assistance where necessary. Li thought that such teaching required a lot from teachers. For example, they had to clearly know their students' learning abilities and problems and then adjust the difficulty level of the problems to the point where students can promote their further learning. Li's words indicate that his ideal teaching style is learner-centred, while his actual teaching, as he reported, is more traditional.

Li stated that 'the present teaching system' requires that mathematics be taught to all. The public often evaluates schools' teaching quality based on students' achievements in the *Gaokao*. As such, teachers often focus their teaching on students' performances and pay the most attention to students who have low academic ability or bad learning habits so as to elevate the average *Gaokao* scores. As he had been teaching in such a system for ten years, he was accustomed to delivering teaching that emphasised content, particularly that required for students' good examination performance.

This was the first time Li had worked as a mentor, so he could only refer his mentoring to his own experiences as a mentee. This led him to consider mentoring to be a simple matter whereby the mentee learned from the mentor by observing his/her teaching; he also thought that this was the traditional form of mentoring in China. Based on his own experiences as a mentee, however, he regarded such mentoring as unnecessary; he believed that teachers' learning relies primarily on the teachers themselves. His experiences as a mentor led Li to believe that his mentoring might have been more effective if he had initiated more discussions with Tommy, as such interactions created learning opportunities.

The interactions showed Li that Tommy was more familiar with some aspects of teaching, such as teaching content and procedure, than others. They also indicated that Tommy spent less time preparing each lesson and was less nervous in the classroom. However, Li pointed out that Tommy spoke softly in classroom teaching which he had told Tommy many times.

6.5.5   Summary and Discussion

In the end stage, based on his teacher-controlled classroom instruction and his focus on subdividing solution procedures for types of questions commonly found on the *Gaokao*, Tommy still appeared to deliver straightforward, teacher-centred teaching with the aim of promoting students' performance. Tommy tended to hold a mix of traditional and non-traditional beliefs about the nature of mathematics. He considered mathematics a kind of knowledge that included both mathematical topics and thinking. For mathematics teaching and learning, he differentiated his practical objectives for teaching in school from his long-term teaching goals, which focused on the development of students' thinking for their individual development and were largely learner-centred; however, his short-term, practical objective was to promote students' performance in the *Gaokao*, and Tommy discussed mathematical topics and problem-solving methods from the perspective of examination success.

The section that follows discusses Tommy's learning outcomes and how those outcomes were achieved, based on the above descriptions of his teaching.

6.5.5.1 Tommy's Overall Learning Outcomes

Over the two-year teaching period, Tommy's stated beliefs changed from non-traditional to a blend of traditional and non-traditional. At the beginning stage, he emphasised the dynamic nature of mathematics, and considered it a developmental process. At the end stage, he began to focus on the static nature of mathematics, as revealed by his perception of mathematics as knowledge. Regarding the teaching and learning of mathematics, he focused on the practical objective of promoting students' performance in the *Gaokao* and turned his views on learner-centred teaching into his long-term goal.

Although Tommy held stated beliefs that were close to the non-traditional category at the beginning, he implemented teacher-centred teaching. He mainly relied on his own interpretations of content in delivering his teaching and explained the content in detail to the whole class to ensure the students understood it. In the second year, as his students changed, he adjusted his teaching to a slower, simpler and more straightforward approach that focused more on students' examination performance, which was consistent with his practical teaching objective. He carefully demonstrated how to solve the types of questions commonly asked on examinations and was learning to teach in a more normative way, with his mentors' guidance.

At the beginning, Tommy appeared to have an urgent need to know what content he should teach to his students and how. His two years' teaching and his own reflections revealed that he had acquired a range of knowledge pertaining to teaching based on his practice, such as new interpretations of content from an educator's perspective, different demonstrations to cater to students of different academic ability and what should be taken into consideration when conducting normative teaching.

6.5.5.2 The Way to Achieve His Learning Outcomes

Unlike Doris and Jerry, Tommy appeared to have little teaching-related experience that could contribute to the original beliefs or knowledge he brought to his practice. However, given his extensive learning experience with both secondary and advanced mathematics, he was confident in his comprehensive understanding of the subject. This may have contributed to his dynamic view of mathematics, but he lacked the required effective pedagogical knowledge to put that view into practice. Thus, there existed a significant contradiction between his stated and enacted beliefs. Tommy practised a rather teacher-centred teaching, with an emphasis on the descriptive explanation of content. Moreover, he himself recognised that he had little knowledge of how to make his teaching more interesting, such as what content he should teach students or effective ways to mobilise their thinking. Tommy attributed this to his lack of teaching experience, and sought to improve through mentoring situated in the school context.

From the investigation of Tommy's teaching over the two years, it seems the school environment supported his consistently practising teacher-centred, content-focused, or even performance-oriented teaching, regardless of whether such teaching may have originated from his lack of pedagogical knowledge. Such environmental influences on his teaching may have directly impacted his acquisition of knowledge and changes in his beliefs or teaching behaviours. For example, given the unified teaching schedule and the emphasis on students' examination performance, Tommy was more likely to acquire knowledge relating to the curriculum and to emphasise the importance of students' achievements. In addition, interactions were evident among Tommy's teaching beliefs, knowledge and teaching practice. For instance, when teaching the Xinjiang class, he had to consider the students' natures and take into consideration their weaker knowledge base and relatively low academic ability. Given his successful implementation of performance-oriented teaching, his belief in practical teaching objectives may also have increased. To sum up, the interactions within his individual teacher system or between Tommy and the school environment indicate a similar process to Doris's professional learning that focused on school mathematics teaching (Figure 4.2, Chapter 4).

In addition to his lack of effective knowledge, the environment likely played an important role in the change in Tommy's stated beliefs from non-traditional to mixed. Firstly, he received no supports from the school when he appeared to lack an effective pedagogical approach to conveying his non-traditional beliefs about the nature of mathematics, such as when he introduced Russell's paradox to students in the beginning stage. Li's mentoring of Tommy was limited to allowing him to observe the more experienced teacher's classroom teaching; Tommy's actual pedagogical learning relied on Tommy himself. However, the teaching that Li demonstrated to Tommy was focused on examinations and emphasised normative teaching, basic rules for teaching and understanding of the centralised curriculum rather than Li's own ideas about teaching. Without the necessary supports, Tommy's lack of related knowledge resulted in ineffective teaching relating to his non-traditional views and the subsequent weakening of those beliefs (Buehl & Beck, 2014). In particular, teaching Xinjiang class students during his second teaching year led Tommy to focus more on students' performance, which was contrary to his original beliefs that emphasised students' interests and efforts. Thus, Tommy's two-year professional learning period reveals a progressive decline in his non-traditional beliefs about the nature of mathematics as well as reciprocal relationships among those beliefs and his knowledge, practice and important environmental influences (see Figure 6.1).

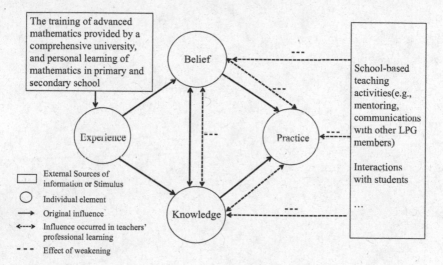

Figure 6.1 The process of decline in the non-traditional beliefs Tommy held at the beginning

The Mentorship For Doris, Jerry, And Tommy

7

7.1 Introduction

The three preceding chapters detailed the three novice teachers' professional learning processes, in which mentorship played an important role. This chapter synthesises the three cases to examine the opportunities and constraints that various mentorships present to novice mathematics teachers' professional learning.

7.2 Seven Mentors for Doris, Jerry and Tommy

As mentioned in 3.3.3.2, the induction programme organised by Shanghai's educational administration mandates mentoring for novice teachers, with the aim of promoting their professional development. The induction programme's guidelines specify that mentoring support must be provided for new teachers across their first year of teaching, and the three schools at which Doris, Tommy and Jerry worked also provided mentoring for them in their second teaching year. The mentoring occurs mainly within LPGs, so that the novice teachers can observe their mentors' teaching on the same mathematical topics and content and initiate discussions with their mentors. Most mentors also assumed the role of LPG head, which allowed the mentees to keep abreast of the entire LPG's teaching progress.

From September 2013 to May 2015, seven mentors were assigned to the three novice teachers, six of whom were mentioned in the three preceding chapters. Ms Zheng was assigned as Doris's mentor for mathematics teaching in the first teaching year; Mr Zhao was originally assigned as Doris's mentor for classroom

X. Lu, *Novice Mathematics Teachers' Professional Learning*, Perspektiven der Mathematikdidaktik, https://doi.org/10.1007/978-3-658-37236-1_7

management in the first year and also took over the mentoring of mathematics teaching in the second year. Doris considered Mr Zhao to be a real '*shifu*' (Master) for her, since she had learned a lot from him.

Mr Qian was assigned as Jerry's mentor for both classroom management and mathematics teaching in the first teaching year. In the second year, Sun took over Qian's role since Qian did not work in the same LPG as Jerry. The mentoring situation for Tommy is more complex. Mr Zhou was assigned as Tommy's mentor for mathematics teaching during his one-semester internship in School C and subsequently continued in the role. However, according to Tommy, Mr Zhou was unable to observe his classroom teaching and discuss it with him, since Zhou was busy teaching mathematics to Grade 12 students and preparing them for the *Gaokao*, while Tommy taught Grade 10 mathematics. However, the requirement that the district implement the Shanghai educational administration's induction programme meant that Tommy should have had a senior teacher as his mentor, and this is why Mr Zhou continued in the role. However, Mr Li executed the major mentor's role, as he worked in the same LPG as Tommy during the first teaching year. Ms Wu became Tommy's mentor during the second teaching year when Tommy was appointed to teach Xinjiang students in a different grade to Li's.

The three novice teachers usually experienced one-to-one mentorship during their early teaching careers. Information about their mentorships is summarised in Table 7.1.

Table 7.1 One-to-one mentorship for the three novice teachers

The mentee	Time period	The mentor
Doris	The first teaching year	Ms Zheng, mentor for mathematics teaching Mr Zhao, mentor for classroom management
	The second teaching year	Mr Zhao, mentor for mathematics teaching and classroom management
Jerry	The first teaching year	Mr Qian, mentor for mathematics teaching and classroom management
	The second teaching year	Mr Sun, mentor for mathematics teaching and classroom management

(continued)

Table 7.1 (continued)

The mentee	Time period	The mentor
Tommy	The first teaching year	Mr Zhou, mentor for mathematics teaching and classroom management Mr Li, mentor for mathematics teaching and classroom management
	The second teaching year	Mr Zhou, mentor for mathematics teaching and classroom management Mr Wu, mentor for mathematics teaching and classroom management

7.3 The Types of Mentorship

7.3.1 Mentor's and Mentee's Input

The mentor's role in the mentorship was analysed based on the interview data from the mentees according to four categories: situation, observing, observed and discussion or general reflection (3.5.3). As such, three of the four categories of mentorship were identified between the three novice teachers and their mentors: (1) mentor-active and mentee-active, (2) mentee-active but mentor-inactive, and (3) mentor-inactive and mentee-inactive.

7.3.1.1 Mentor-Active and Mentee-Active

Mentor-active and mentee-active mentorship was recognised when specific teaching practice was conducted, as in the cases of Doris and Tommy. Mr Zhao was actively involved in the mentorship when Doris integrated mathematical history into teaching. He came to observe Doris's classroom teaching on his own initiative and subsequently discussed her teaching with her. In particular, Zhao actively supported Doris in integrating mathematical history into her teaching in public lessons by encouraging her and helping her to prepare her lessons. Zhao also found that he appreciated Doris's efforts to integrate mathematical history into teaching, which was considered a kind of non-utilitarian teaching practice.

Tommy reported that Mr Li also actively discussed solutions to difficult mathematical problems with him when Li and Tommy shared the teaching for an optional mathematical competition class. As Tommy had graduated from a prestigious comprehensive university with a master's degree in mathematics, he was perceived as having a better understanding than those who had graduated with degrees in mathematics education.

7.3.1.2 Mentor-Inactive and Mentee-Active

Most of the time, the mentorship between the three novice teachers and their mentors was recognised as mentor-inactive and mentee-active, including the mentorships between Doris and Zhao, Jerry and Qian, Jerry and Sun, and Tommy and Li.

Doris reported that she regularly observed Mr Zhao's mathematics teaching in the classroom, focusing on the examples that Zhao typically used in class and his different approaches to introducing mathematical topics. She usually initiated discussions with Zhao to obtain a comprehensive understanding of Zhao's selection of mathematical examples and approaches to introducing topics. When she experienced problems in her own teaching preparation and delivery, she also asked for Zhao's comments and he usually offered direct suggestions.

Jerry frequently observed Mr Qian's classroom teaching during his first teaching year to enhance his teaching experience. He reported that he usually focused on Qian's presentation of mathematical topics, including the examples he used and his organisation of the demonstration. Jerry tried to implement Qian's method of teaching in his own classes. He said that he wished to learn from Qian how to teach using 'a clear structure of mathematical knowledge' and 'good skills in classroom management so that the students would listen carefully (to the teacher)'. In accordance the Shanghai induction programme's requirements, Qian was obliged to observe Jerry's teaching several times. Qian asked Jerry to give him his lesson plan beforehand and, based on the lesson plan and his observations, Qian usually discussed Jerry's problems with him, pointing out areas for improvement based on his opinion.

Jerry also observed Sun's classroom teaching, but his observations became less frequent over time. He noted that Sun emphasised students' performance in the *Gaokao*. As such, he focused on observing Sun's teaching of topics relevant to the *Gaokao*. Without the constraints of the Shanghai induction programme during Jerry's second teaching year, Sun seldom actively observed Jerry's teaching or initiated discussions with Jerry about his teaching.

Tommy relied heavily on Li during his first teaching year, owing to his inexperience. He said that he was used to teaching mathematical topics based on his own experiences as a student. However, he found what he learnt from his own learning experience could not be directly applied in teaching, particularly owing to the diversity of students' abilities. Thus, he usually sought to emulate Li's teaching by observing Li in the classroom one day before he was due to teach. He believed that Li had a comprehensive understanding of the mathematical topics, and by observing Li's teaching, he also learned Li's way of understanding mathematical topics, such as the meanings and applications of mathematical concepts.

With these, he could understand the proper structure to present the mathematical topics in his own class. He also learned from Li about students' possible misconceptions of mathematical topics and which mathematical examples could help reveal these misconceptions. Mr Li also followed the Shanghai induction programme's instructions to observe Tommy's classroom teaching. According to Tommy, Li usually pointed out his problems after the observation, focusing on Tommy's teaching style. He usually asked Tommy questions such as 'Why did you teach in this way?', 'Why did you use this sequence?', and 'What were your intentions when teaching in this way?' to help Tommy reflect on his teaching.

7.3.1.3 Mentor-Inactive and Mentee-Inactive

The mentorships between Doris and Zheng, Tommy and Zhou, and Tommy and Wu can be characterised as mentor-inactive and mentee-inactive.

According to Doris, all teachers in the same LPG opened their classrooms to her for observation, so she was not wholly reliant on Zheng's mentorship, which largely took the form of classroom observation. Moreover, thanks to her year's teaching experience in the west of China, Doris was confident in her teaching abilities and was able to conduct teaching independently during her first teaching year at School A. For her, reflecting on her own teaching was more important than observing others at that time.

As mentioned above, Mr Zhou was appointed Tommy's mentor as a senior teacher. However, when Tommy was teaching Grade 10, Zhou taught Grade 12 students to help them prepare for the national examination. As such, Zhou was busy, with little time to observe Tommy's classroom teaching, while his teaching was not useful for Tommy to observe. However, Tommy was pleased to receive Zhou's teaching plans for mathematics teaching throughout Grades 10 to 12, which gave him some spare time as he did not need to prepare the lessons himself. He could use Zhou's plans directly or make some revisions for his classes. During the second teaching year, Tommy was busy with his own teaching and did not have much time to observe Ms Wu's classroom teaching. Neither could Wu come to observe his. They had few opportunities to communicate with one another except for sharing the teaching schedule to maintain the same pace. Moreover, Tommy thought Wu's teaching style was different from his own which emphasised more on students' comprehensive understanding of mathematical contents. As required by School C's mentorship system, Mr Zhou and Ms Wu came to observe Tommy's classroom teaching but only infrequently.

7.3.2 Mentee's Self-reflection and Attitudes Towards Mentorship

7.3.2.1 Mentorship Considered Supportive

The mentorships between Doris and Zhao, Jerry and Qian, and Tommy and Li were considered to be supportive by the three novice teachers.

Doris was highly self-reflective during her early teaching career. She said that she observed each lesson with a specific aim, such as 'stealing the mathematical examples Mr Zhao used'. Doris believed that the mentee should be more active in the mentoring process, posing questions and initiating discussions. She said that she preferred to teach in her own way first and then observe Mr Zhao's classroom instructions and discuss potential problems in her own teaching with him. She found that her discussions with Zhao helped her reflect on her own teaching, particularly because Mr Zhao's teaching approach was consistent with hers—namely, encouraging student initiative.

Doris had a generally positive attitude towards Mr Zhao's mentoring. She found Zhao's classroom instructions interesting and was able to develop her own teaching skills by reflecting on her observations, such as different approaches to the introduction of various mathematical topics. Moreover, she appreciated Zhao's encouragement to solve the problems she faced during the teaching practice. She found Zhao particularly supportive in one instance when she prepared a public lesson:

> He talked with me about my lesson plan design, the entire classroom teaching procedure, each detail of the teaching, the design of blackboard writing and where I should be cautious when presenting almost everything.

She said she 'felt very good' when she improved on the aspects that her mentor emphasised.

Jerry considered Qian's mentorship to be supportive, although Qian was required by the induction programme to observe Jerry's classroom teaching and initiate discussion with Jerry based on his observations as well as opening his classroom to Jerry to freely observe. Jerry valued this mentorship during his first teaching year. He said that he had known very little about teaching at that time, and he was keen to learn from experienced teachers. He believed Qian's mentoring came at the right time.

Tommy reflected on the differences between his classroom teaching and Mr Li's. He described Li's strengths, including his 'well-structured' approach and ability to 'lead students to understand by themselves'. By contrast, he thought his own teaching was 'unimpressive', 'too talkative', saying that the students usually 'appeared to be bored'. He attributed these differences to his lack of teaching experience:

> I think it is a disadvantage that I did not graduate from a normal university. That's why I cannot behave like other teachers. What I can follow is only my own experience of learning specific mathematical content.

Tommy thus relied on following Li's teaching to improve his own. He stated, 'without observing his teaching, I don't know how to teach (specific mathematical topics), such as the demonstration structure and in which sequence the mathematical topics in one lesson should be demonstrated'. Moreover, Tommy thought he had a similar teaching style to Mr Li, paying attention to students' comprehensive understanding of the mathematical content, and he found that Li was successful in this regard. Tommy thus continued to observe Mr Li's teaching during the second teaching year when Li was no longer his mentor to see 'how he dealt with the (the teaching of) mathematical topics'.

7.3.2.2 Mentorship Considered Non-Supportive or Non-Inductive

Jerry found Sun's mentoring to be non-supportive. Doris did not express any clear opinion towards Tian's support as a mentor, and Tommy also offered no strong opinions on Zhou and Wu's supportiveness as mentors.

Jerry considered Sun's mentoring to be non-supportive when Sun replaced Qian as his mentor during the second teaching year. He said that it was because he had 'already known how to teach' from having observed Qian's teaching and he 'had too much work to do' to observe Sun's teaching frequently. He tried to conduct his teaching independently during the second teaching year.

As the mentorship between Doris and Zheng was mentor-inactive and mentee-inactive, few interactions took place between them. Doris did not express a clear opinion of Zheng's mentoring, and therefore, it is not possible to determine whether she evaluated the mentoring as supportive.

Tommy also had fewer interactions with Zhou and Wu during his two years of teaching, so no clear evidence can be presented with respect to the mentors' supportiveness.

7.4   The Opportunities and Constraints of Mentorship

7.4.1   General Mentorship Types: A Summary

The above detailed descriptions of the three novice teachers' perceptions of the mentorship they experienced demonstrate that various mentorship categories can be empirically distinguished by differentiating four mentoring styles wherein either both the mentor and mentee are inactive or active or with one active and the other inactive. Furthermore, by integrating the mentees' attitudes towards the mentorship's supportiveness, four general types may be generated from the above-described seven one-to-one mentorship types—demonstrative, supportive, collaborative and unnecessary (as summarised in Table 7.2).

In demonstrative mentorship, the mentee actively observes the mentor's classroom teaching and following the mentor's teaching style, but the mentor is inactive. This mentorship type appears useful for the mentee during the initial stages of mentoring and is thus usually recognised as supportive by the mentee. Jerry and Qian's mentorship relationship during Jerry's first teaching year exemplifies this type, as does the mentorship between Tommy and Li. When the mentee believes that they have acquired the mentor's teaching style, such mentorships may become non-supportive from the mentee's perspective, as exemplified by Jerry and Sun's relationship.

In supportive mentorship, the mentee is active and the mentor is primarily inactive but is supportive when necessary. Doris and Zhao's relationship is mainly characterised by this type. Doris conducted her own teaching independently but could approach Zhao and discuss any problems she might have encountered.

In collaborative mentorship, both the mentee and mentor are active and supportive from the mentee's perspective. The mentorships between Doris and Zhao and Tommy and Li can exemplify this type in specific instances. Zhao became more active and the mentorship developed a more reciprocal character when Doris began to integrate mathematical history into her teaching. Zhao believed that he had learned from Doris about teaching practice. Li also became active when he shared teaching work with Tommy in an optional mathematical competition class and actively discussed solutions to difficult mathematical problems with Tommy.

Unnecessary mentorship is observed when both the mentee and mentor are inactive and the mentee has no clear opinion of their mentoring experience in terms of whether it is supportive or non-supportive. This mentorship type is exemplified by the mentorships between Doris and Zheng, Tommy and Zhou, Tommy and Wu, and Jerry and Sun. A mentorship may become unnecessary when, for example, the mentee prefers to conduct teaching on their own (e.g., the mentorship between Doris and Zheng), the mentor does not teach the same group of students as the mentee (e.g., the mentorship between Tommy and Zhou), the mentee perceives the mentor as having a different teaching style to theirs (e.g., the mentorship between Tommy and Wu) or the mentee finds it useless to observe and follow the mentor's teaching as they have already become acquainted with teaching practice (e.g., the mentorship between Jerry and Sun).

7.4.2 The Nature of the Four Types of Mentorship: A Discussion

As mentioned above, the seven mentorships between the three novice teachers and their seven mentors during their first two teaching years may be categorised into four types (Table 7.2) based on assessment of the mentee's and mentor's input and the mentorship's supportiveness from the mentee's perspective.

Table 7.2 The types of mentorship for the three novice teachers

Mentorship	Mentorship type
Doris and Zheng	Unnecessary
Doris and Zhao	Mainly supportive and sometimes collaborative
Jerry and Qian	Demonstrative
Jerry and Sun	Unnecessary
Tommy and Zhou	Unnecessary
Tommy and Li	Mainly demonstrative and sometimes collaborative
Tommy and Wu	Unnecessary

Three of the four mentorship types are considered to be useful for the mentee: demonstrative, supportive and collaborative mentorship. In demonstrative and supportive mentorship, the mentee is active and the mentor is inactive, while collaborative mentorship requires that both the mentor and mentee are active.

More specifically, demonstrative mentorship allows mentees who have little pedagogical experience to learn directly from experienced teachers (e.g., the mentorships between Jerry and Qian and Tommy and Li). In this case, the mentee can adapt to teaching quickly, which is particularly relevant within a centralised curriculum and may be considered an advantage. However, this mentorship style may deprive the mentee of the opportunity to explore their own teaching independently. This can lead to the abandonment of mentoring when the mentee is confident in teaching independently and has the impression that no support is needed anymore, as exemplified by Jerry's case. Therefore, mentorship of this type may not be sustainable. It shares several similarities with the supervision mentorship style used during the probation phase in New South Wales in Australia (Kemmis et al., 2014).

Supportive mentorship allows the mentor to assist (rather than supervise) the mentee in learning to teach. This mentorship type requires that the mentee already have some pedagogical qualifications from the outset in addition to the mentor's availability to assist the mentee. It also likely requires good timing: for example, when Doris was busy with her own teaching and eager to reflect more on her teaching at the beginning, she was unwilling to seek support from Ms Zheng. Supportive mentorship shares commonalities with support mentoring in Sweden, as reported by Kemmis et al. (2014).

Collaborative mentorship requires a joint understanding between the mentor and the mentee based on similar teaching ideas or styles so that the mentor can actively work with the mentee. In such mentorships, the mentee can be encouraged to learn from the mentor and develop their own specific teaching practice based on intensive conversations with the mentor; meanwhile, the mentor may also learn about new teaching approaches from their mentee. For example, Doris appreciated Zhao's encouragement and support, which she considered to have a positive influence on her determination to incorporate mathematical history into her teaching. This mentorship type may promote collaboration between novice teachers and experienced teachers and thus contribute to a positive atmosphere of teacher professional development within the teacher community.

A single mentor–mentee dyad may include more than one mentorship type— for example, the mentorships between Doris and Zhao and between Tommy and Li—depending on the context, different teaching content or approaches. The mentorship may be or become unnecessary for various reasons. The seven mentorships in this study provide four case studies (as summarised in Table 7.2).

7.4.3 Conclusions

The above analyses on the mentorship of the three novice teachers indicate that the mentor's active input will be highly appreciated by the mentee and may be regarded as a decisive factor for the mentorship's sustainability. However, most of the seven mentorship groups were characterised by inactive mentors. This may be because the induction programme was newly developed and not widely established, although it requires supporting active mentorship (e.g., by stipulating the frequency and times at which the mentor observes the mentee's classroom teaching). Moreover, it may also be attributed to the cultural characteristics of mentorship in China (going beyond the mentorship for teaching but for every profession) that emphasise the necessity of the mentee's own input. As an old saying goes, 'the mentor can open the door, but you must enter by yourself' (that is, the mentor may open their classroom for you to observe, but you must learn of your own accord). However, the importance of the mentor's active input should also be better recognised with the aim of simultaneously promoting novice teachers' professional learning and experienced teachers' professional development.

This chapter has analysed the different opportunities and challenges affecting novice mathematics teachers' professional learning—in particular, three mentorship styles: demonstrative, supportive and collaborative. It also has revealed the factors that contribute to unnecessary mentorship in the context of China. However, owing to the limited number of cases included, this study may have missed other potential mentorship types and factors determining unnecessary mentorship. For a more comprehensive understanding of mentorship, more cases of mentees with different backgrounds should be analysed.

Discussion And Conclusion Of The Study

8

Chapters 4 to 6 individually presented Doris's, Jerry's and Tommy's experiences of professional learning during the initial two years of teaching in the schools in which they were situated. This chapter first summarises and discusses the findings from the analyses of the three professional learning cases to answer the two research questions. It then outlines the implications of the study. The final section offers several suggestions for further research.

8.1 Discussion of Findings in the Study

As mentioned in Chapters 2 and 3, this study aimed to investigate the professional learning process that novice mathematics teachers experience during the early stages of their careers in situated contexts by asking two general research questions:

1. What learning outcomes could the novice mathematics teachers obtain during the observed two years of teaching, particularly in terms of teacher beliefs, teacher knowledge and teaching practice?
2. How was professional learning implemented to achieve the learning outcomes?

This section will respond to these questions by summarising and discussing the findings from all three cases.

X. Lu, *Novice Mathematics Teachers' Professional Learning*, Perspektiven der Mathematikdidaktik, https://doi.org/10.1007/978-3-658-37236-1_8

8.1.1 Learning Outcomes

Answers to the first question constitute answers to the three sub-questions detailed in Chapter 3 concerning the original beliefs and knowledge that the teachers held when they began teaching, the features of their teaching practice at different stages and the features that remained consistent or changed.

8.1.1.1 Teacher Beliefs

When the three novice teachers began teaching, their beliefs about the nature of mathematics were close to non-traditional; for example, Tommy considered mathematics to be a developmental process. The three teachers all emphasised the importance of students' interests and the development of students' mathematical thinking in mathematics learning and teaching. After the two-year teaching period, their beliefs tended to be more traditional but mixed, emphasising the static nature of mathematical topics as well as the connections between topics. Regarding mathematics teaching and learning, they paid particular attention to students' performances in the large-scale public examination, the *Gaokao* (i.e., the college entrance examination).

This shift in teacher beliefs (from non-traditional to traditional) may not be surprising in the context of China. Ma et al. (2006) noted that Chinese mathematics teachers typically tend towards traditional teaching practices, even if they hold relatively non-traditional beliefs, owing to the high pressure relating to national or large-scale public examinations; researchers have recognised that teaching practice can effect a change in teacher beliefs (e.g., Bandura, 1997; Guskey, 1986; Lumpe, Czerniak, Haney, & Beltyukova, 2012; Tschannen-Moran & McMaster, 2009). Moreover, studies in the Chinese context have commonly found that most Chinese mathematics teachers consider mathematics to be static, emphasise the connections between mathematical facts and incline towards teacher-centred beliefs about mathematics teaching and learning (e.g. An et al., 2006; Cai, 2007; Wang & Cai, 2007; Wong, 2002). In this study, these findings were triangulated by the investigation of the two participating mentors (Doris's mentor Teacher Zhao and Tommy's mentor Teacher Li) in this study. In this sense, the mixture of traditional and non-traditional beliefs about the nature of mathematics and teacher-centred beliefs about mathematical teaching and learning appear to be collectively held by teachers in the context of China. The three novice teachers who brought different learning experiences to bear on their teaching aligned their own beliefs with these collective beliefs over the two years; this implies that the environment may play an important role in their professional learning.

Based on the three cases and their mentors' statements at the end of the two-year period, it also became clear that their ideas around mathematics teaching differed from their perceptions of the mathematics that they actually taught. They differentiated the two teaching types from the content that they taught, their teaching goals and the learners. For example, Doris distinguished scientific mathematics from school mathematics; Tommy differentiated between teaching for 'implicit' and 'explicit' knowledge; Jerry, Tommy and Zhao separated examination-oriented from interest-oriented teaching; and Li argued that mathematics teaching should ideally target several learners, rather than all. Their conceptualisations of mathematics teaching and learning appeared closer to non-traditional beliefs, while their beliefs about school teaching tended to be more traditional.

Chen and Leung (2014) also observed conflicting pedagogical beliefs among Chinese mathematics teachers. Based on a review of previous studies conducted in China, they asserted that teachers recognise the importance of 'student-centeredness' but that such 'student-centeredness' usually refers to the general needs of student groups rather than individual students' specific needs. This is not the same as the contradictory beliefs held by the teachers in this study. For instance, as Zhao stated, mathematics teaching and learning, in his opinion, should be 'non-utilitarian' and centred on the development of mathematics, while the actual teaching of mathematics in schools is 'utilitarian', with an emphasis on students' examination performance. All three teachers reported in their interviews that they were obliged to implement examination-oriented teaching owing to high pressure from school leaders, parents and students, who overemphasise the importance of the *Gaokao*. According to Zhao, experienced teachers in China seem to have grown accustomed to such examination-oriented teaching; as such, did the novice teachers make any effort to retain their original non-traditional approach? Doris did: over the two years, she insisted on promoting students' interest in learning mathematics by integrating mathematics history and culture into her teaching, although she also emphasised the content that she taught and considered the teachers' role to be that of an explainer.

Mathematical history is a pedagogical tool that offers new perspectives and insights into materials preparation for lessons as well as a goal for learning (Jankvist, 2009). In mainland China, the importance of mathematical history in mathematics education has been increasingly recognised. The High School Mathematics Curriculum Standard (experimental) (Ministry of Education of China, 2003) considers 'reflecting the cultural value of mathematics' to be fundamental and recommends that mathematical culture be integrated into upper secondary school mathematics curricula. Chinese researchers (e.g., Wang, X., 2013) have advocated promoting the history and pedagogy of mathematics (HPM) in real teaching situations. Wang believed that by integrating history into teaching, teachers could form special teaching patterns and promote critical thinking with respect to textbooks as well as an awareness of the potential to expand on the material presented in textbooks (Wang, X., 2013). Doris appears to have incorporated HPM in her teaching. She reported having been encouraged and trained to integrate history into teaching by her master's degree programme; this implies that relevant supports and training are necessary to promote teachers' innovative teaching ideas.

The above discussion appears consistent with the findings of several comparative studies between mainland China and other regions (East and West)—that large class sizes, mandated curricula and large-scale public examinations (Chen & Leung, 2014; Ma et al., 2006) or the effects of culture (An et al., 2006; Leung, 1992) explain why most Chinese teachers' beliefs about teaching evolve into mixed beliefs. However, through a longitudinal case study approach, this study provides a detailed description of the mechanisms by which this change occurs from a teacher learning perspective with respect to the interactions among participating teachers' beliefs, knowledge and teaching practices as well as interactions between the individual teachers and their environments.

8.1.1.2 Teaching Practice

An investigation of the beginning, medium and end stages of their first two years of teaching found that all three novice mathematics teachers consistently delivered teacher-centred, content-focused and performance-oriented teaching, regardless of the teacher beliefs they held. For example, Tommy focused largely on teaching content and delivered traditional teaching that relied on his own explanation of the content, although in interviews he appeared to hold non-traditional beliefs about the nature of mathematics and learner-centred beliefs about mathematics teaching and learning.

To present the characteristics of these novice teachers' teaching, this study provided detailed descriptions of the activities they prepared, implemented and reflected on in different stages of their classroom teaching, many of which have been presented in other studies of mathematics teaching in Chinese classrooms. Reviewing a substantial body of literature, Fan, Miao and Mok (2014) highlighed nine aspects of these features: planning lessons systematically; emphasising two basics; whole-class teaching and instruction; teaching with variation; teacher-student interactions and engagement; assigning and marking homework frequently; using textbooks with deep understanding; structured instruction; and implementing change in response to curriculum innovations. The three teacher cases presented in this study focused on the deep understanding and structured interpretation of teaching content by referring to various teaching materials, including textbooks (both teacher and student versions), supplementary teaching materials bought in book market or downloaded from the internet, and other teachers' lesson plans when preparing lessons. They planned the lessons following the unified teaching schedule provided by their respective LPGs to ensure that they maintained the same teaching pace as the other teachers. In implementing the lessons, they prioritised clear explanation of the contents and employed numerous easily answered questions to ensure structured instruction and invite the entire class to interact with them.

Their instruction was teacher-controlled, as the teachers consistently retained the authority to initiate questions and determine answers' correctness. In particular, they emphasised students' understanding of the content and their performance in daily homework and on progress tests. Relying on their review of students' mistakes when doing related exercises, the teachers could identify problems in students' understanding and then act to consolidate that understanding. After class, they typically focused on any problems they had encountered in implementing their lessons, such as the smoothness of the demonstrations. During the two years, they reflected less frequently on teaching because they appeared to have become familiar with teaching through their teaching practice or were too busy with their daily teaching.

These common features among the three teachers were not formed in the same way, probably owing to the different experiences they brought to their teaching. Among them, Doris appears to have been the first to exhibit these features in her relatively independent teaching shortly after she started teaching in School A. She received teacher training in both her bachelor's and master's programmes and had experienced one year of teaching as well. On the other hand, Tommy expressed anxiety about school teaching at the beginning stage because of his lack of pedagogical training. He and Jerry relied heavily on demonstrative mentorship by observing and following experienced teachers' (i.e., their mentors') classroom instructions in the first teaching semester.

Their different experiences led to the three teachers having different learning experiences over the two years, but their resulting learning outcomes, in terms of teaching practice, had much in common. They all emphasised the importance of content and examination, which have been recognised as traditional teaching methods by studies in China and other countries in East Asia over the past two decades (e.g., Biggs, 1996; Huang & Leung, 2005; Leung, 2001). An et al. (2006) explained that in China, teacher-centred teaching is considered a heuristic method by which teachers can inspire and promote students to think deeply and to learn actively. They also attributed the use of such teaching to large class sizes and cultural influences, again implying that context matters in teacher professional learning.

In China, teachers are usually required to acquire the basic skills to implement classroom teaching in a normative way. This phenomenon was seen in the cases of Doris and Tommy across the two years. They were particularly required by their mentors to, for example, clearly express themselves to students in class, ensure their blackboard demonstrations were neat and structured and design their instruction purposively (i.e., to know clearly how and why to teach the content). According to local documents, these were traditionally considered essential skills for teachers, which people generalised as 'three writing and one language' (*san-zi-yi-hua*, 三字一话), including brush writing, pen writing, blackboard writing and Mandarin (Wan, 1999). It appears that the mentors' guidance in this study emphasised the novice teachers' acquisition of these traditional basic skills; Zhao, for example, opined that acquiring these skills could help Doris in presenting herself as an accomplished teacher.

However, local researchers have recognised that these basic skills should be assigned new meanings given the development of curriculum and teacher teaching (e.g., Xu, 2004). Yang (2015) suggested that teachers' basic skills could be categorised according to the various aspects of their teaching delivery, including the skills to prepare lessons, the use of various teaching patterns, the use of the blackboard and multimedia, the skills for conducting educational research and so on. Kong (2006) also claimed that, in addition to the basic skills (e.g., the skills to organise teaching, use of blackboard and students' feedback), teachers should also acquire other skills to promote professional development, such as proper guidance from experts and more communication with teacher colleagues. In this study, different types of mentorship provided different opportunities and challenges for the three novice teachers' professional learning (detailed analyses in Chapter 7), which appears to heavily influence their professional learning processes and outcomes.

In particular, Doris appeared to exhibit some differences in her teaching practice. She endeavoured to integrate mathematical history into her teaching to promote students' interest in mathematics while still implementing teacher-centred and content-focused teaching. This phenomenon is not unique to mathematics teaching in China. To understand East Asian students' consistent outstanding achievements in international assessments, researchers have investigated the mechanisms of mathematics teaching in East Asia (e.g., Huang & Leung, 2005; Mok, 2006); Huang and Leung (2005) claimed that teacher-dominated classroom teaching could also include elements of student-centeredness, and that students could also actively generate knowledge, even under teacher-controlled instruction. Doris reported in the interviews that by integrating mathematical history into teaching, she aimed to promote students' interest in and exploration of mathematics. Her learning of teaching also focused on this, and she tried multiple approaches to integrate history into the teaching of various mathematical topics. In the second semester, she provided students with a mathematical problem from historical materials to solve outside of class time. Later, in the third semester, she drew on the origin of *analytic geometry* to emphasise the connection between mathematical content and the usefulness of mathematics. Finally, in the final semester, she even changed a lesson's teaching goal from 'teaching formulae for volumes of solids' to 'introducing Zu Geng's principle and its application', in which she aimed to teach students the derivation of the formulae. Doris's teaching practice indicated that she embedded some elements of 'student-centeredness' into her teacher-centred approach.

To sum up, the three novice teachers' teaching practices tended to conform with what are considered to be common features of Chinese mathematics teachers' teaching—i.e., teacher-centred, content-focused and examination-oriented— even though the three teachers brought unique experiences to their teaching careers and originally held more non-traditional teacher beliefs. In addition, most of the mentorship for the novice teachers (e.g., demonstrative and supportive mentorship) focused on normative ways of teaching, centred on traditional concepts of teachers' basic skills in China. The above suggests that their respective environments played a more important role in the teachers' adopting relative traditional teaching practices than did their previous experiences. Doris was something of an exception in that she integrated elements of 'student-centeredness' into her teaching, which also led to the collaborative mentorship; however, her teaching was still based on a primarily teacher-centred model.

8.1.1.3 Teacher Knowledge

During the two-year teaching practice, the three novice teachers always strove to have a comprehensive interpretation or deep understanding of the content so that they could explain it clearly to students. Consequently, they usually studied available supplementary teaching materials when preparing lessons. According to Leung (2001), Confucian culture has a deeply rooted tradition that a mathematics teacher should be an expert or learned figure (a scholar) in mathematics; this image of a scholar-teacher has profoundly influenced education in East Asia to the extent that subject matter knowledge alone is not sufficient. Thus, the teachers were learning knowledge for teaching on the way that they practiced teaching. For example, when Doris implemented her pre-prepared lesson plans in her classroom teaching, she found she had underestimated her students' academic abilities; she then studied school-developed student exercise books to better understand the features of School A's students.

Face with the environmental pressure associated with the *Gaokao*, all three paid increased attention to the mandated curriculum and unified progress examinations. In her self-reflection on her two years of learning to teach, Doris in particular pointed out her comprehensive understanding of the mathematics curriculum content for Grade 10 and 11 students. Jerry and Tommy emphasised the key contents and types of questions commonly seen on examinations; they relied heavily on the authority of experts, including adherence to the textbook or to lesson plans developed by experienced teachers, and followed the classroom instruction methods that experienced teachers implemented. In the task of 'discussing mathematical problems', all three paid particular attention to mathematical knowledge that frequently occurs in the examination.

As the above description illustrates, the novice teachers acquired and accumulated knowledge of content, students and curriculum to complement their teaching practice. As discussed in 8.1.1.2, their teaching practice was strongly influenced by the environment; thus, the environment was essential to their learning of teacher knowledge.

Moreover, their previous learning experiences appeared to influence their learning of knowledge. Studies of pre-service mathematics teachers in China indicate that they have acquired a profound understanding of both subject matter and pedagogical knowledge (Fan, Miao, & Mok, 2014). Influenced by China's long history of centrally designed curricula, textbooks and examinations (Wong, Han, & Lee, 2004), Chinese mathematics teachers are trained to deliver curricula as designed (Zhang & Wong, 2014) and usually focus on students' performance in large-scale examinations (Ma et al., 2006). This evidence suggests that the three teachers should have acquired knowledge for teaching that focused on mathematical content, students, curriculum and examinations in the pre-service teacher education or induction programmes they experienced. However, their learning of teacher knowledge was likely impacted by the training's rigour and intensity. Unlike Doris, for example, Jerry and Tommy—who had little or no teacher training—had no specific way of dealing with their lack of knowledge of students in the initial stage except to gain experience from their teaching practice or by observing their mentors' teaching. Later in the two years, they understood their students better through teacher–student interactions. Sufficient teacher training is necessary for teachers attain competence in the initial stages of their careers.

In Doris's case, her learning experience not only contributed to her competence in daily teaching but also helped her attain the knowledge of how to integrate mathematical history into teaching, including related historical knowledge, instructional cases that show how to integrate historical knowledge into her teaching and knowledge of how to search for and research related packets of knowledge for further related learning. By contrast, Tommy's lack of effective pedagogical knowledge may have led to his failure to practise his non-traditional beliefs.

Generally, the experiences that the novice teachers brought to their teaching may have equipped them with knowledge for a particular teaching approach (e.g., Doris's integration of history into teaching) and impacted their teacher knowledge learning process, such as learning what kind of knowledge to use and when and how to use it. However, their knowledge learning outcomes were focused on teaching that emphasises content, curriculum and especially examination, due to environmental influences.

8.1.2 The Process of the Two-year Professional Learning

The above section discussed the three teachers' learning outcomes during the two-year period, in terms of teacher beliefs, knowledge and teaching practice. Many of these findings corroborate those of earlier studies carried out in the context of China, but the present study places particular emphasis on the mechanisms by which these novice mathematics teachers' professional learning was accomplished. From the above discussions, both consistencies and inconsistencies among teacher beliefs, knowledge and teaching practiced existed in the two years. Woolfolk Hoy, Hoy and Davis (2009) suggested that dissonance is necessary for teachers to experience change; thus, from a cognitive perspective on learning theory, this section will discuss the interactions of the elements in the individual teachers' systems, including their experience, knowledge, beliefs and practice. It will also discuss the influences of the environment in which the teachers' professional learning was situated, which reveals the situated learning perspective involved in this study.

Based on the detailed investigation of the three cases (Chapters 4 to 6), a generic model (see Figure 8.1) that involves the influential individual elements (experience, beliefs, knowledge and practice) and environmental factors is established to show the professional learning process the three teachers experienced. The process begins from the box on the left side of Figure 8.1, which shows the related sources or stimuli (e.g., pre-service teacher education or induction programmes) that may contribute to the experiences the cases brought to their teaching. It continues with the kinds of original beliefs and knowledge shaped by such experiences and how they influenced their teaching practice. The process particularly reveals reciprocal relationships among the individual elements and shows the impacts of teaching practice on teacher beliefs and knowledge. Moreover, the environment in which the teachers worked is involved in the process through its influences on teacher beliefs, knowledge and practice with sources (e.g., supplemented teaching materials) and stimuli (e.g., mentorship).

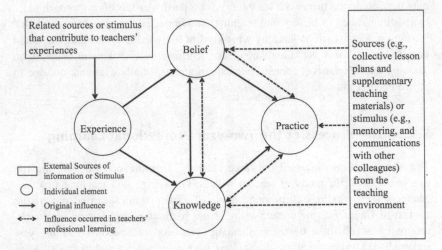

Figure 8.1 A generic model showing the process of professional learning

Using this model, the kind of professional learning carried out by the teachers can be summarised from the stories presented in this study. Three significant findings:

1) All three novice teachers seemed to experience a professional learning process for teacher-centred, content-focused and performance-oriented teaching, in which a mix of traditional and non-traditional beliefs was sustained or established and their acquisition or accumulation of knowledge was guided by the centralised curriculum system (in particular, its large-scale examinations).

2) Doris, at the same time, also engaged in learning for the practice of integrating mathematical history into teaching, which reflected her belief in promoting students' interest and exploration in mathematics and allowed her to consolidate and acquire related knowledge.

3) With the mixed beliefs confirmed, it appeared that Tommy rejected learning in favour of the non-traditional beliefs he brought to bear on his teaching career.

To present the complex and dynamic processes that the three teachers adopted or rejected in their learning, we will first discuss how teacher beliefs, knowledge and teaching practice interacted or combined to impact other elements. We will then discuss the significant influences of the experiences the teachers brought into their teaching as well as the environment in which their teaching was situated.

8.1.2.1 The Interactions Among Individual teachers' beliefs, Knowledge and Practice

It was found in this study that the interactions between teacher knowledge, beliefs and practice could constitute the novice mathematics teachers' own orientation to learning, which appeared consistent with the complex conceptualisation proposed by Opfer and Pedder (2011). For example, the beliefs held by teachers could impact their decisions on teaching (Raths, 2001). Considering the teacher's role to be that of an explainer and keeping her emphasis on teaching contents in mind, Doris conducted teacher-centred and content-focused classroom teaching. In addition, as she thought it important to promote students' interest, and integrating the history and culture of mathematics into teaching could be an effective way to achieve this, she made efforts to do so. Moreover, beliefs also influence teachers' willingness to learn for teaching (Opfer & Pedder, 2011); to complete her integration of history into teaching, Doris set out to acquire the necessary knowledge, including knowledge relating to the history of the teaching contents and specific ways to embed mathematical history into her classroom instruction.

However, teachers may not be conscious of their beliefs or the effects thereof (Furinghetti, 1997; Furinghetti & Pehkonen, 2002). Tommy, for instance, seemed unaware of his teacher-centred and content-focused approach when he stated his beliefs about the teaching of mathematics though he enacted it in his teaching practice at the beginning stage. It is not unusual to perceive inconsistencies between teachers' beliefs and practice (Cross, 2009), particularly among novice teachers (Raymond, 1997). However, various factors may lead to such inconsistencies, including individual teachers' psychological constructions (such as teacher identity and teacher efficacy), pedagogical knowledge, and environmental factors such as school culture and curriculum features (Clarke & Hollingsworth, 2002; Fennema & Franke, 1992; Raymond, 1997). Teacher knowledge, another important element in the interactions between teacher beliefs and practice, is usually considered to work interactively to influence teaching practice (An, Kulm, & Wu, 2004). This study largely agrees with Ernest's (1989) view that beliefs play the role of regulator between knowledge and teaching behaviours. With their mixed beliefs, the three teachers conducted their related learning to better understand their students' characteristics and the content required for examinations to implement their teacher-centred, content-focused and examination-oriented teaching.

However, if the dissonance between their beliefs, practices and knowledge is too large, teachers may dismiss new ideas as inappropriate to their situations (Coburn, 2001). In the case of Tommy, it seemed that he abandoned his non-traditional view very early in the beginning stage due to his lack of related pedagogical knowledge, and his teaching practice became more traditional. The self-efficacy of teachers' teaching is regarded as an important factor in enhancing, diminishing or sustaining their beliefs (Buehl & Beck, 2014). Tommy turned to performance-oriented beliefs when he found that his method of teaching consisting of detailed explanation of content was ineffective. Moreover, students' active responses to Doris's integration of mathematical history into teaching may have increased her corresponding belief in the promotion of students' interest through such integration. When Jerry found his close attention to students' practice could indeed improve their learning performance, he tended to have more confidence in content-focused and performance-oriented beliefs about teaching and dismissed his previous emphasis on students' interest. As a result, the teacher-centred, content-focused and performance-oriented teaching was consistently implemented. As long as teachers change, their orientation towards learning also changes (Opfer & Pedder, 2011).

The above discussion suggests that teacher beliefs, knowledge and practice interact and combine to influence teachers' teaching and their pedagogical learning. Nonetheless, if the dissonance among the elements is too great, it can promote teachers' rejection of learning. Thus, the intensity of interactions also matters (Opfer & Pedder, 2011). Teacher beliefs, knowledge and teaching practice are all influenced by the experiences the teachers bring to their teaching and by the external sources and stimuli in their situated teaching environment; as such, their experiences and the environment also influence teachers' professional learning.

8.1.2.2 The Influence of the Novice Teachers' Experiences

It has been widely recognised that experiences contribute to the pedagogical beliefs and knowledge teachers that bring to their teaching and affect their teaching practice (Novak & Knowles, 1992; Richardson, 1996; Richardson, 2003). First, it was found that Doris and Jerry, who brought mixed beliefs to their teaching, experienced teaching-related training in teacher preparation institutions—normal universities. This made us consider what kinds of training prospective teachers receive in Chinese teacher-preparation institutions. In their review of studies in pre-service training programmes in China, Fan, Miao and Mok (2014) observed significant similarities between the programmes provided by institutions in different regions of China. Li's (2002) two-stage study also showed that mathematics teaching courses generally focus on student teachers' deep understanding and the thorough integration of mathematical subject knowledge, student cognitive development and pedagogical principles. Through comparative studies in teacher-preparation programmes in China and Western countries (e.g., the US), Fan et al. (2014) found that Chinese student teachers receive more training on mathematical content, particularly advanced mathematics topics, and therefore have more comprehensive subject matter knowledge. Zhang and Wong (2014) also believed that Chinese prospective teachers are trained to deliver the curriculum as designed. Thus, it appears that the training that teachers receive encourages them to focus on mathematical content and conduct their teaching in adherence to the curriculum, which is consistent with the beliefs that Doris and Jerry held and likely led to their teacher-centred and content-focused teaching in the beginning stage.

Comparison of the three teachers' teaching at the beginning of the two years indicates that their teacher training did impact their teaching. Doris, who experienced two training programmes (her bachelor's and master's degree programmes) from a normal university, was able to engage in daily teaching independently when she started teaching. Without having experienced any teacher training, Tommy exhibited a lack of confidence in his teaching at the beginning. He and Jerry relied heavily on imitating their mentors' teachings. However, Jerry had undergone teacher training in his normal university bachelor's programme and even expressed his appreciation for the training, particularly the opportunity to teach in real school contexts (practicum), from which he realised the differences between teacher and student perspectives on contents or problems. Doris pointed out, in her interviews, that a practicum will not be very effective if its duration is too short or the interval between the practicum and beginning teaching is too long. Thus, she complained that her undergraduate training was useless. Given that it did not obviously result in effective teacher training outcomes for Jerry and that Doris explicitly called it ineffective, should we reflect on and invest greater effort in improving existing teacher training? Chinese researchers (e.g., Wang, J., 2013) have asserted that the training provided by teacher-preparation institutions should be the beginning and foundation of teachers' professional development, during which teachers should be equipped with the knowledge and skills necessary for them to develop and thrive in the teaching profession.

Doris appreciated the teacher training provided by her master's programme, from which she learned the importance of promoting students' interest in mathematics in teaching, that integrating the history and culture of mathematics could be an effective way to achieve this and the necessary knowledge to do so in the two-year teaching period; she even engaged in new learning to promote her teaching practice. On the other hand, Tommy, who emphasised the dynamic aspects of mathematics and learners' efforts in learning how to teach mathematics, could not effectively conduct his learner-centred teaching owing to his lack of related training. Based on his interviews, his non-traditional beliefs likely stemmed from his long-term experience in learning mathematics (in particular, advanced mathematics), which gave him robust confidence in his mathematical knowledge and his comprehensive understanding of scientific mathematics. Both Doris's promotion of students' interest through the integration of history into her teaching and Tommy's learner-centred approach likely originated from their experiences (teacher-training experience or learning mathematics experience). However, these ideas required relevant training to enable the teachers to acquire the knowledge necessary to implement them.

In addition, it was found that Doris's innovative idea was still implemented through teacher-centred and content-focused teaching, which she considered 'the basic teaching' required by the environment. Thanks to her year of teaching experience in China's underdeveloped Western region, Doris appeared to understand basic teaching better, including such rules as controlling each lesson's content capacity and catering to students' diverse academic abilities. Thus, she could independently conduct her teaching at the beginning stage, including preparing her teaching with the use of available materials and flexibly adjusting it in consideration of her students' requirements and characteristics. This implies that her successful innovative teaching ideas would not have been easily implemented if they had differed significantly from the teaching required by the environment.

To summarise, the following findings demonstrate the influences of the experiences the novice teachers brought to their teaching careers:

- The novice teachers' pedagogical beliefs and knowledge were likely to be influenced by the training provided by their teacher preparation institutions. However, this training seemed to direct them to conduct teacher-centred and content-focused teaching, which is consistent with the mixed beliefs transmitted by training in the context of China. In addition, the training provided by bachelor's programmes from normal universities seems not to be as effective as expected, from the individual teachers' perception.

- The teachers could form innovative teaching ideas simply by learning the subject of mathematics or experiencing special teacher training aimed at such ideas; however, to practise those innovative ideas, teachers may need to be equipped with the necessary knowledge.
- It also appears that teacher-centred and content-focused teaching is popularised in and required by the teachers' situated teaching environments. Innovative teaching practice would not easily succeed if it completely contradicted the teaching that was favoured by the environment.

8.1.2.3 The Influences of the Teachers' Situated Environment

From the above discussion, it appears that the teachers' learning had been influenced by the context that all Chinese mathematics teachers shared prior to commencing teaching. However, in this study, we mainly focused on the likely influences of their professional learning environments after they had begun teaching (in particular, the schools in which they worked). Based on the two-year investigation and particularly the individual teachers' perceptions, it appears that two types of influence exerted by the environment can affect teachers' knowledge, beliefs and teaching practice.

Two forms of environmental influence

First, the environment commonly provided several resources to support teaching, including school-developed student exercise books, supplementary teaching materials that the teachers bought in the market or downloaded off the internet and lesson plans developed by experienced teachers or LPGs. The investigation of the three novice teachers' experiences suggested that their use of these sources likely affected their teaching practice and acquisition of pedagogical knowledge and influenced their related beliefs.

These sources were mainly used in the preparation and implementation of lessons as references for the teachers as they designed their teaching plans and selected exercises for their students. By studying these sources, the teachers could improve their understanding of the content they taught, how to teach the content, or even their students' characteristics. However, it is also suggested that they may differ in their use of the same category of sources. For instance, the unified students' exercise books (either school-developed or available on the market) were typically the source of student exercises; among the three teachers, only Doris also used exercise books developed by her school's LPGs to learn what content was most appropriate for teaching her school's students.

In their use of the lesson plans provided by experienced teachers or LPGs, Tommy and Jerry also appeared to place more trust in the authority of experts. They tended to use the ready-made lesson plans with only minor modifications, such as excluding some mathematical examples, and often paid greater attention to understanding the plans so as to implement them correctly. Initially, they used to try to organise their teaching by themselves through their own interpretation of the content or study of the teaching materials. However, they believed that their instruction was ineffective and typically elected to follow the examples they perceived in more experienced teachers' classroom teaching. Their reflections indicate that they did not actually know why their lesson designs were ineffective; they generally attributed it to their lack of familiarity with the students or their teaching inexperience and were inclined to place their trust in the more experienced teachers around them.

Doris also found that her original teaching designs were inappropriate; however, she explicitly attributed this to her undervaluation of her students' academic ability and then turned to school-developed student exercises to learn more about her students. Later, Doris also used the collective lesson plans developed by other upper secondary schools or even directly used ready-made lesson plans downloaded from the internet when she was too busy to prepare her own lessons. However, she emphasised that she had her own designs for the lessons, sometimes including the integration of mathematical history. She stated that she could learn from the collective lesson plans about what aspects experienced teachers considered the key points of a given lesson or used their lesson plans to save time preparing lessons when time was limited. It is possible that the three teachers knew or were able to learn about other teachers' interpretations of the content and their teaching organisation by using or referring to these lesson plans.

The documents the teachers used imply that most teaching materials in China, including school-developed or supplementary materials available on the market and shared lesson plans, usually adhere to the mandated curriculum (e.g., Cai, 2012; Mo & He, 2007), and that the essential aim of these teaching materials is the promotion of students' examination performance through mechanical training (e.g., Cai, 2012; Zhang, 2012). Zhang (2012) found that the exercises in such materials were in traditional and monotonous paper format rather than oral or comprehensive practices. Local documents also highlighted the features of what were considered to be appropriate supplementary teaching materials, such as keeping pace with the textbooks, explaining related mathematical problems in detail, thoroughly analysing the concepts required by the curriculum and including as many question types as possible (Editorial department of a local journal for the mathematics and science education in secondary schools, 2008). Mo and He

(2007) also noted that the development of lesson plans in China must usually be based on the curriculum and textbook to strictly follow written forms and include well-organised and structured teaching objectives, key and difficult points and a clear statement of teaching procedure.

As such, the use of lesson plans may focus teachers' teaching on the delivery of a well-designed centralised curriculum and the promotion of student performance; however, this may also restrict teaching diversity (Clune, 1993). In particular, ready-made lesson plans may deprive teachers of opportunities for learning if they use them as rigid standards rather than reflecting on whether they suit the teaching context and making adaptions as necessary. Chinese local researchers have recognised that 'take-ism' (拿来主义) is inappropriate for the diversity of teachers' professional development, and preparing lessons collectively should focus on learning among teachers, rather than on unifying their lesson plans (e.g., Li & Zhao, 2011; Wang, 2006; Xu, 2010).

Second, the novice teachers reported that the school environment provided many opportunities for them to learn from other teachers through mentorship and collective teaching activities. These interactions could also influence their practice, knowledge and beliefs in the two-year teaching period, in various ways.

Mentorship usually assigns one or two experienced teachers to provide individual novice teachers with instruction for teaching. In this study, all of the assigned mentors were in the same schools and usually the same LPGs as the mentees. The mentees were responsible for availing of mentorship opportunities, while mentors periodically observed the mentees' classroom teaching and provided comments as required by the induction programme. Jerry and Tommy initially appeared to rely considerably on their mentors as a means of learning about daily teaching practices. Almost every day, they observed their mentors' classroom instruction to learn from the more experienced teachers' interpretations of the content and how they taught it; they then incorporated what they had learned into their own teaching. By comparing their teaching with that of their mentors, they could also recognise the weaknesses in their teaching, implying that mentoring had significant influences on the teachers' daily teaching practices and knowledge and likely also influenced their teaching beliefs. For example, mentoring led Jerry to consider strict classroom discipline an effective way of teaching after he observed Qian's good classroom teaching results from strict management.

Later, when they found they had acquired some structure within which to implement their daily teaching, they observed their mentors' teaching less frequently; however, their interactions with colleagues tended to be different because their situations changed. In the second year, Jerry was assigned another mentor, Sun, who was also a member of his LPG. Sun organised weekly teacher meetings, required teachers to collectively prepare lessons and increased the entire group's attention to preparing for the *Gaokao*. Jerry appeared to be heavily influenced by Sun's emphasis on the examination, and he came to regard it—rather than his students' interest—as his teaching objective. In his daily teaching, he often emphasised the key content required for the examination and had students practice extensively in preparation for it. Tommy was assigned, in the second year, to teach another group of students (the *Xinjiang* class) who had much lower academic ability and a weaker knowledge base than the students he had taught in the first year. The mentorship provided by the school still gave him the opportunity to learn from Wu about how she taught such students, although he considered it less necessary.

The influence of her interactions with other teachers on Doris's teaching appears to have differed from that experienced by Jerry and Tommy. At the beginning, Doris said that she did not spend much time learning from colleagues but independently implemented her teaching, with emphases on learning the contents and on students. This indicated that her knowledge and beliefs may have been be sufficient to help her cope with teaching independently. Later, after becoming more familiar with daily teaching and her LPG, she found that interactions with her colleagues (in particular her mentor, Teacher Zhao) contributed significantly to her pedagogical learning. She thought it enriched her knowledge when she observed others' teaching and discussed teaching issues with them. According to Doris, the relatively lower pressure the LPG placed on her regarding students' performance provided her with the opportunity to implement her own ideas with respect to promoting students' interest by integrating mathematical history and culture into her teaching. Teacher Zhao also encouraged her ideas and provided necessary support for her practice. It was also commonly found in all three cases that their mentors usually required them to deliver more normative classroom teaching, including strict requirements for language use, writing and other behaviours (e.g., standing straight) when teaching, as well as the need to follow a purposive instructional design that adhered to the curriculum. As mentioned in 8.1.1.2, such mentoring focused Doris and Jerry on the acquisition of traditional basic skills for teachers in the context of China; new curriculum development and teacher learning skills, such as the integration of technology in teaching and communications and interactions with other teacher colleagues, were seldom promoted by their mentors. In this sense, the mentoring provided by the environment seemed too traditional to enhance novice teachers' innovative teaching practices.

Essence of the influences provided by the environment: the norms of the environment
Although the three teachers interacted with other teachers in different ways, it was commonly found that their interactions usually focused on teacher-centred, content-focused and performance-oriented teaching and normative teaching behaviours in class. It appears from the findings on the beliefs held by the three teachers and their mentors, as well as findings in other studies in the context of China (as discussed in 8.1.1), that teachers collectively held mixed beliefs about the nature of mathematics and teacher-centred beliefs about the teaching and learning of mathematics, which was enacted in their teaching practice. They appeared to share the collective belief that students' performance in the examinations, particularly large-scale public examinations (i.e., *Gaokao*), was important. They usually used students' academic achievements to evaluate their teaching, considered students' understanding of the content an effective way to achieve students' good performance, and relied on teacher-dominated instruction and the repeated honing of mathematical skills to achieve student understanding. This reveals the teachers' collective beliefs about efficacy, which contributed to the establishment of environmental norms (Tschannen-Moran, Salloum, & Goddard, 2014).

From the environmental norms, situated group members could learn to behave in accordance with the overall group and evaluate each other's compliance (Bandura, 1989). According to Tschannen-Moran et al. (2014), norms promote interactions between personal, behavioural and environmental factors and reflect the 'triadic reciprocal causation' suggested by Bandura. It is believed that a school's functioning relies heavily on its academic and social norms (Bandura, 1997; Goddard & Goddard, 2001; Tschannen-Moran, Salloum, & Goddard, 2014). Tschannen-Moran, Salloum and Goddard pointed out two powerful norms governing teachers' behaviours: academic pressure and teacher professionalism. According to Hoy, Hannum and Tschannen-Moran (1998), academic pressure refers to collective environmental (e.g., school) beliefs, with a clear emphasis on academic achievement. It includes shared confidence in students' abilities and the belief that all students can attain high academic standards. Through these, the environment emphasises academic achievement and affects the normative behaviours of the involved teachers, who may, in turn, contribute to the high-pressure environment.

The present study's findings imply that there existed high academic pressure in the three schools in which the teachers worked and that their professional learning over the two years was strongly influenced by that academic pressure in both direct and indirect ways. All three teachers reported that they were under high academic pressure from the environment. For Jerry and Tommy in particular, it seemed that Schools B and C had a higher degree of academic pressure. Jerry admitted that he neglected the development of students' interest due to the school's indifference to it. Tommy also recognised that 'the most practical objective' for his teaching by the end of the two-year teaching was 'letting students enter universities' with high achievements in the *Gaokao*. Even Doris, who thought that the environment (e.g. the LPG) she was in did not particularly emphasise students' academic achievement, still considered the teaching of curriculum contents on a regular teaching schedule and the promotion of students' good unified examination achievements to be 'the basic teaching tasks' she had to complete on a daily basis. In addition, as mentioned above, the sources the teachers used to implement their teaching were likely to have a direct influence on their teaching and focused it on the curriculum and examinations. Since the advent of imperial China's examination system centuries ago, it has been widely believed that examinations are of great importance in Chinese education (Wu, 2012). Thus, academic pressure seems to be a strong nationwide norm that could govern teachers' behaviour in China (Ma et al., 2006).

Through productive collaboration, de-privatised teaching practice and reflection as well as a collective focus on students' learning, it is believed that teachers may be socialised into the shared norms of the teaching profession, teachers' beliefs and behaviours (Seashore Louis, Kruse, & Marks, 1996; Tschannen-Moran, Salloum, & Goddard, 2014). According to Tschannen-Moran, Salloum and Goddard (2014), teacher professionalism refers to '[teachers'] perceptions that their colleagues take their work seriously, demonstrate a high level of commitment, and go beyond minimum expectations to meet the needs of students' (p. 7). A high degree of teacher professionalism usually reflects teachers' trust in their colleagues' competence and expertise and a resultant willingness to work cooperatively with them (Hoy, Hannum, & Tschannen-Moran, 1998). Such collaboration among teachers may enhance different perspectives on solving the problems teachers face in teaching (Hoy & Sweetland, 2001) and thereby support student learning. Fan, Miao and Mok (2014) observed that China has developed a coherent and institutionalised in-service professional development system for teachers to collaborate and interact with each other through collaborative activities, including school-based teaching research activities, the development of public lessons, in-service training, and apprenticeship practices.

It appears from the findings in this study that the novice teachers frequently interacted with experienced teachers, particularly through mentoring, and likely benefited from such products of collaborative teaching as unified teaching schedules, school-developed exercise books and collective lesson plans. Teachers tend to demonstrate their trust by interacting with other teachers and using their products. The teachers in the present study usually kept pace with the unified teaching schedules, learned about teaching knowledge through use of the teaching materials, obeyed their mentors' instructions (particularly regarding normative classroom teaching behaviours) and so on. Based on the collective beliefs emphasising performance-oriented teaching, teacher professionalism in China may positively correlate with student achievement, which is consistent with the findings of Tschannen-Moran et al. (2006). Thus, it is hypothesised from the three teachers' high level of trust in their colleagues and its correlation with students' achievement that a high degree of teacher professionalism could also exist in the teachers' situated environment and affect their professional learning.

Professional learning in the normative environment
It appears from the above discussions, that two powerful norms probably affected the novice teachers' professional learning: academic pressure and teacher professionalism. Consequently, their professional learning tended to focus on teacher-centred, content-focused and performance-oriented teaching, which is consistent with other teachers in the same environment, even though one (Tommy) held non-traditional beliefs about the nature of mathematics and emphasised the students' own efforts in the learning and teaching of mathematics in the beginning. Doris' case seems to be exceptional in that she insisted on promoting students' interest in her teaching over the two years; however, while it is an exception, it is not unexpected, as both studies and the curriculum in China advocate the integration of mathematical history in teaching. In addition, it was also found in this study that the participating teachers (both novice and experienced) still held 'non-utilitarian' and student-centred views of mathematics teaching, although they adhered to the teacher-centred and examination-oriented practices required by the environment. Therefore, Doris experienced a relatively low degree of academic pressure from the school, and she was supported by her mentor in promoting students' interest by integrating mathematical history into her teaching.

However, Doris emphasised that her efforts to integrate mathematical history into her teaching were based on her implementation of 'the basic teaching' and the result of a high degree of academic pressure. This made the practice difficult, and she could only use her spare time to engage in learning for her teaching practice. Doris's promotion of her students' interest remained embedded in content-focused teaching practices approved by environmental norms; by comparison, Tommy's student-centred teaching for the non-traditional view of mathematics he held in the beginning stage appeared to differ considerably from the collective beliefs shared by the teachers in his environment. This implies that environmental norms only allow teachers to adopt new learning that largely aligns with the belief systems from which the norms stem, in addition to pedagogical learning within the systems.

8.2   Significance of the Study

Based on the above discussions, this study's theoretical and practical significance lies in its research methodology, its contribution to mathematics teachers' professional learning or development, and its insights into the teaching and learning of novice teachers in specific contexts, considering both individual teacher differences and the impact of the environment.

8.2.1 Theoretical Significances

First, the study contributes to research methods for analysing the complex and dynamic theory of teacher professional learning. It demonstrates how a case study approach may be used to provide an in-depth description of what and how novice teachers obtain during the pedagogical learning process and illustrates individual differences in their professional learning. It employs classroom observation and interviews in a longitudinal investigation of the teachers' beliefs, knowledge and teaching practices and then uses the investigation to illustrate in detail the interactions among teacher experience, beliefs, knowledge and practice in the individual teachers' subsystems as well as the interactions between those subsystems and the teachers' situated environments (especially their schools) in detail. With the completed case examples, the study enriches the complex conceptualisation of teacher professional learning, which is—referring to the theoretical framework of this study (Figure 2.1 in Section 2.5, Chapter 2)—a dynamic process combining multiple strands of theory pertaining to teacher professional development, teaching and learning, teacher change and organisational learning.

Second, with respect to novice teachers, the study has corroborated and/or extended our knowledge of early-stage teacher expertise. In addition to the interactions among the beliefs, knowledge and practice of individual teachers, the influences of teacher experience and the environmental norms in teachers' situated contexts should be taken more wholly into consideration. The study illustrates the important role that teachers' training-related experience plays in novice teachers' teaching and pedagogical learning, particularly the intensity of that training. It also provides empirical data to support the assertion that environmental norms stemming from the collective beliefs of members situated in the environment are vital factors in novice teachers' professional learning. As discussed in 8.1.2.3, norms probably impact teachers' adoption and rejection of new learning through their use of resources provided by the environment and interactions with other teachers in the same teaching environment.

Third, this study also contributes to our understanding of mathematics teacher education and of mathematics teaching and learning in China. As indicated in 8.1.1 and 8.1.2, its findings imply considerable commonality in the novice teachers' learning outcomes and the professional learning they conducted, which seems consistent with the common features of mathematics teaching in Chinese classrooms that other researchers believe are based on deep-rooted cultural values and paradigms (Leung, 2001). The study provides empirical data from the perspective of individual teachers to offer hypotheses about how culture impacts teacher professional learning in China, particularly for novice teachers. The widespread and long-term influences of China's centralised curriculum (Wong, Han, & Lee, 2004) and examination system, which can be traced back to imperial China (Wu, 2012), suggest teachers' collective beliefs in and emphases on teaching contents and students' performance as integral elements of a normative school environment (Tschannen-Moran, Salloum, & Goddard, 2014). Given the teaching practices in such normative environments and their interactions with others in the environment, novice teachers tend to behave normatively and develop or maintain beliefs consistent with the collective ones.

8.2.2   Practical Significances

The study contains descriptive and analytic data regarding the ways in which teacher professional learning changes or is sustained. In particular, from the perspective of novice mathematics teachers in specific contexts, it offers implications for in-service teacher education, particularly for novice teachers in China or similar contexts.

First, based on individual differences in the practice and learning of teaching in the early career stages, the importance of pre-service training for teaching may be detected. Without it, teachers may not have sufficient pedagogical knowledge to complete the basic teaching tasks required by schools or to practise their own ideas. In addition, according to the teachers' teaching performance in the initial stages of their teaching careers and their impressions of the training they received, it also suggests that the teacher training provided by the bachelor's programme in normal universities in China warrants some improvement. This study also implies that specific training focused on teachers' implementation of new ideas in teaching is important so as to equip teachers with necessary ideas and knowledge. However, owing to a lack of empirical data, it is difficult to determine whether it is more effective to deliver such training during pre- or in-service teacher education.

Second, the study also provides novice mathematics teachers with information regarding the supports and constraints they will likely encounter when teaching in real school contexts and implies that environmental norms may govern their perceptions and behaviours with respect to teaching. Novice teachers should be better informed about their situations so that they can exercise autonomy in their own professional learning and make learning choices that are consistent with environmental norms or that conform with their own ideas of teaching.

Third, the study also has implications for school leaders and policy makers. As illustrated above, teachers' professional learning is likely influenced by norms stemming from the collective beliefs of teachers in the environment. It is also believed that school leaders and policy issues play an important role in fostering collective beliefs and shared norms (e.g., Goddard & Salloum, 2011; Tschannen-Moran, Salloum, & Goddard, 2014). This study's descriptive and analytic data provide information on the different needs of individual teachers. It may contribute to the construction of school cultures that are composed of tacit assumptions and beliefs and that govern the beliefs and behaviours of teachers situated in the school (Schein, 2006).

8.3   Limitations of the Study

The study has several limitations in terms of its research methodology, such as its selection of informants and data collection time points, emphasis on qualitative interpretation and its case study approach, which limits the generalisability of its findings. The study selected novice teachers who had obtained master's degrees and who were employed by key upper secondary schools, because schools tend to employ more highly educated teachers, and because key schools are perceived as having better facilities to enhance teaching and learning and attract more attention from the public. However, such selectivity will inevitably lead to a lack of diversity in the professional learning of novice teachers. Moreover, since 'grade levels present very different contexts' (Fullan & Hargreaves, 1992, p. 6), different findings may emerge if the investigation to include teachers working in primary or junior secondary schools or in Grade 12, in which students often face a significantly different instructional reality, as preparations for the *Gaokao* are far more important and occupy far more class time than in the Grade 10 and 11 classes that the novice teachers taught. Moreover, given contextual differences, this study's findings or implications may not be as applicable to teacher professional learning in systems with decentralised curricula or other different features, such as economic factors. As a primarily qualitative study, the researchers' insights are the main instrument of analysis (Lincoln & Guba, 1985). The researcher herself, who has limited research experience, is inevitably implicated in the study through her own biases and interpretations.

In addition, the study has explicitly focused on the professional learning of novice teachers at selected time points in the initial two years of their careers from the perspectives of individual teachers. Considering the complexity of the development of teacher expertise, a two-year period may be too short, and significant learning phenomena will inevitably be missed. According to a survey conducted by Lian (2008) of some 3,000 teachers in mainland China, based on their psychological features, novice teachers are those who have four to six years of teaching experience. A richer and better understanding of their professional learning might be obtained if a longer or follow-up investigation were to be conducted. Moreover, although the importance of teachers' experience was recognised in the study, little attention was paid to the mechanisms by which such experience exerts an influence. The study also did not focus on students or organisational learning, which are, of course, crucial elements in teacher professional learning. Its investigation of individual teachers placed too much emphasis on the teachers' own perceptions. However, fortunately, the researcher will have the opportunity to access the same research site to conduct further research in the future. The rich hypotheses generated by this study will thus be tested with larger samples in the future, considering different teachers' backgrounds and working environments. The section that follows offers some suggestions in this regard.

8.4   Suggestions for Further Research

As Fan, Miao and Mok (2014) pointed out, international research in Chinese mathematics education is still in its infancy, and much more remains to be done. Based on the issues that could not be resolved in the present study but that are still of importance, the following research areas or questions are suggested for further investigation, particularly in the context of China:

1) How new learning for novice mathematics teachers might be developed or sustained, including the strategies that could be used and when those strategies would be most effectively implemented.
2) How a wider sample of teachers, especially those in different stages of teacher expertise, might adopt or reject the programmes with the aim of teacher professional development and what professional learning outcomes they may have.
3) Whether an organisation also performs its own learning when interacting with teachers, including what sort of organisational learning could be conducted and how and how it might impact individual teachers' development.
4) The student perspective on teacher professional learning, including how they might positively engage in teachers' learning, and whether they might simultaneously obtain learning opportunities for themselves.
5) The situation for teachers in different contexts (e.g., under a decentralised curriculum system), including the theoretical implications when comparing findings in different contexts for teacher professional learning/development.

8.5   Envoi

The originality of this study lies mainly in its comprehensive analysis of teacher professional learning on both the individual and the school/organisational levels, through a longitudinal case study approach that includes rich interview and classroom observation data. It contributes to the literature by analysing how teachers' experiences, beliefs, knowledge and practices interact to produce various learning outcomes and the role the environment plays in novice mathematics teachers' implementation of their teaching in the context of Shanghai, China. Its key finding is that the environment in which the novice teachers are situated plays an essential role in their professional learning, particularly through its norms; that the teachers brought different levels and types of experience to their teaching, which impacted the initiation and trajectory of their learning rather than their learning outcomes; and that, through the effect of its norms, the environment will likely allow teachers to engage only in new learning that does not contradict the collective beliefs and teaching practices from which the norms stem. Finally, the study's significance and implications are mainly reflected in its confirmation and development of the complex and dynamic theory of teacher professional learning/development, in particular the implication that teacher professional development is heavily influenced by environmental norms, so that the cultural appropriateness of teachers' learning in given contexts can be investigated.

References

Adelman, C., Jenkins, D., & Kemmis, S. (1976). Re-thinking case study: notes from the second Cambridge Conference. *Cambridge journal of education, 6*(3), 139–150.

Ambrose, R., Clement, L., Philipp, R., & Chauvot, J. (2004). Assessing prospective elementary school teachers' beliefs about mathematics and mathematics learning: Rationale and development of a constructed-response-format beliefs uurvey. *School Science and Mathematics, 104*(2), 56–69.

An, S., Kulm, G., & Wu, Z. (2004). The pedagogical content knowledge of middle school, mathematics teachers in China and the US. *Journal of Mathematics Teacher Education, 7*(2), 145–172.

An, S., Kulm, G., Wu, Z., Ma, F., & Wang, L. (2006). The impact of cultural differences on middle school mathematics teachers' beliefs in the US and China. In F. K. S. Leung, G. K. D., & F. J. Lopez-Real (Eds.), *Mathematics Education in Different Cultural Traditions- A Comparative Study of East Asia and the West: the 13th ICMI study* (pp. 449–464): Springer Science & Business Media.

Anderson, J. R., Reder, I. M., & Simon, H. A. (1996). Situated learning and education. *Educational researcher, 25*(4), 5–11.

Anderson, J. R., Reder, L. M., & Simon, H. A. (1997). Situative versus cognitive perspectives: Form versus substance. *Educational researcher, 26*(1), 18–21.

Argyris, C. (1993). *Knowledge for action: A guide to overcoming barriers to organizational change.* San Francisco, CA: Jossey-Bass.

Argyris, C., & Schön, D. A. (1978). *Organizational learning: A theory of action perspective.* Reading, MA: Addison-Wesley.

Ary, D., Jacobs, L. C., & Razavieh, A. (2002). *Introduction to research in education* (6th ed.). Belmont, CA: Wadsworth/Thomson Learning.

Avalos, B. (2011). Teacher professional development in Teaching and Teacher Education over ten years. *Teaching and Teacher Education, 27*(1), 10–20.

Ball, D. L. (1988). *Knowledge and reasoning in mathematical pedagogy: Examining what prospective teachers bring to teacher education.* (Unpublished doctoral dissertation), Michigan State University, East Lansing.

Ball, D. L., & Bass, H. (2003). Making mathematics reasonable in school. In J. Kilpatrick, W. G. Martin, & D. Schifter (Eds.), *A research companion to principles and standards for school mathematics* (pp. 27–44). Reston, VA: National Council of Teachers of Mathematic.

Ball, D. L., & Cohen, D. (1999). Developing practice, developing practitioners: Toward a practice-based theory of professional education. In G. Sykes & L. Darling-Hammond (Eds.), *Teaching as the learning profession: Handbook of policy and practice.* (pp. 3–32). San Francisco: Jossey-Bass.

Ball, D. L., Thames, M. H., & Phelps, G. (2008). Content knowledge for teaching what makes it special? *Journal of teacher education, 59*(5), 389–407.

Bandura, A. (1986). *Social foundations of thought and action: A social cognitive theory.* Englewood Cliffs, NJ Prentice Hall.

Bandura, A. (1989). Human agency in social cognitive theory. *American Psychologist, 44*(9), 1175–1184.

Bandura, A. (1997). *Self-efficacy: The exercise of control.* New York: Freeman and Company.

Barth, R. S. (1986). The principal and the profession of teaching. *The Elementary School Journal, 86*(4), 471–492.

Berebitsky, D., Goddard, R. D., Neumerski, C., & Salloum, S. J. (2012). The influence of academic press on students' mathematics and reading achievement. In D. M. F. & F. P. B. (Eds.), *Contemporary challenges confronting school leaders* (pp. 51–72). Charlotte, NC: Information Age.

Berliner, D. C. (1988). *The development of expertise in pedagogy.* Paper presented at the the annual meeting of the American Association of Colleges for Teacher Education, New Orleans, LA. Retrieved on February 2, 2022 from eric.ed.gov/?id=ED298122

Beswick, K. (2005). The beliefs-practice connection in broadly defined contexts. *Mathematics Education Research Journal, 17*(2), 39–68.

Beswick, K. (2007). Teachers' beliefs that matter in secondary mathematics classrooms. *Educational Studies in Mathematics, 65,* 95–120.

Biggs, J. (1996). Western misperceptions of the Confucian-heritage learning culture. *The Chinese learner: Cultural, psychological and contextual influences,* 45–67.

Bishop, A., Seah, W. T., & Chin, C. (2003). Values in mathematics teaching: The hidden persuaders? In A. J. Bishop, K. Clements, C. Keitel, J. Kilpatrick, & F. K. S. Leung (Eds.), *Second international handbook of mathematics education* (pp. 717–765). Dordrecht, Netherlands: Kluwer Academic Publishers.

Blömeke, S., Felbrich, A., Müller, C., Kaiser, G., & Lehmann, R. (2008). Effectiveness of teacher education. *ZDM—Mathematics Education, 40*(5), 719–734.

Bogdan, R., & Biklen, S. K. (1997). *Qualitative research for education.* Needham Heights, MA: Allyn and Bacon.

Borich, G. D. (2000). *Effective teaching methods* (4th ed.). New Jersey: Prentice-Hall, Inc.

Borko, H. (2004). Professional development and teacher learning: Mapping the terrain. *Educational researcher, 33*(8), 3–15.

Buehl, M. M., & Beck, J. S. (2014). The relationship between teachers' beliefs and teachers' practices. In H. Fives & M. G. Gill (Eds.), *International handbook of research on teachers' beliefs (Educational psychology handbook)* (pp. 66–84). New York: Routledge.

Burden, P. R., & Byrd, D. M. (2007). *Methods for effective teaching: Promoting K-12 student understanding.* Boston: Pearson Education, Inc.

Cai, H. (2012). Xiaoben Zuoye: "xiaoben kecheng" kaifa de yansheng [School-based assignment: the extension of the development of "school-based curriculum"]. *Jiaoxue Yuekan (Zhongxue Ban) [Teaching Monthly (Middle School Edition)]* (10), 35–36.

Cai, J. (2007). What is effective mathematics teaching? A study of teachers from Australia, Mainland China, Hong Kong SAR, and the United States. *ZDM—Mathematics Education, 39*(4), 265–270.

Chapman, O. (2002). Belief structure and inservice high school mathematics teacher growth. In G. Leder, E. Pehkonen, & G. Törner (Eds.), *Beliefs: A Hidden Variable in Mathematics Education?* (pp. 177–193). Dordrecht: Kluwer.

Chen, Q., & Leung, F. K. S. (2014). Chinese teachers' mathematics beliefs in the context of curriculum reform. In L. Fan, N.-Y. Wong, J. Cai, & S. Li (Eds.), *How Chinese teach mathematics: Perspectives from insiders* (pp. 529–566). Singapore: World Scientific Publishing Company.

Chen, X., & Li, Y. (2010). Instructional coherence in Chinese mathematics classroom—A case study of lessons on fraction division. *International Journal of Science and Mathematics Education, 8*(4), 711–735.

Chester, M. D., & Beaudin, B. Q. (1996). Efficacy beliefs of newly hired teachers in urban schools. *American Educational Research Journal, 33*(1), 233–257.

Clark, C. M. (1988). Asking the right questions about teacher preparation: Contributions of research on teacher thinking. *Educational researcher, 17*(2), 5–12.

Clarke, D., Emanuelsson, J., Jablonka, E., & Mok, I. A. C. (2006). *Making connections: Comparing mathematics classrooms around the world*. Rotterdam: Sense Publishers.

Clarke, D., & Hollingsworth, H. (1994). *Reconceptualising teacher change*. Paper presented at the Proceedings of the 17th annual conference of the Mathematics Education Research Group of Australasia.

Clarke, D., & Hollingsworth, H. (2002). Elaborating a model of teacher professional growth. *Teaching and Teacher Education, 18*(8), 947–967.

Clarkson, P., & Bishop, A. (1999). *Values and mathematics education*. Paper presented at the 1st Conference of the International Commission for the Study and Improvement of Mathematics Education, University College, Chichester, UK.

Clune, W. H. (1993). The best path to systemic educational policy: Standard/centralized or differentiated/decentralized? *Educational evaluation and policy analysis, 15*(3), 233–254.

Cobb, P., & Bowers, J. (1999). Cognitive and situated learning perspectives in theory and practice. *Educational researcher, 28*(2), 4–15.

Cobb, P., Wood, T., & Yackel, E. (1990). Classrooms as learning environments for teachers and researchers. In R. Davis, C. Maher, & N. Noddings (Eds.), *Constructivist views on the teaching and learning of mathematics* (pp. 125–146). Reston, VA: National Council of Teachers of Mathematics.

Coburn, C. E. (2001). Collective sensemaking about reading: How teachers mediate reading policy in their professional communities. *Educational evaluation and policy analysis, 23*(2), 145–170.

Cohen, L., Manion, L., & Morrison, K. (2007). *Research Methods in Education* (6 ed.). Oxon, New York: Routledge/Falmer.

Cole, A. L. (1989). *Making explicit implicit theories of teaching: Starting points in preservice programs*. Paper presented at the the annual meeting of the American Educational Research Association, San Francisco.

Coleman, J. S. (1985). Schools and the communities they serve. *The Phi Delta Kappan, 66*(8), 527–532.

Coleman, J. S. (1987). Norms as social capital. In G. Radnitzky & P. Bernholz (Eds.), *Economic imperialism: The economic approach applied outside the field of economics* (pp. 133–155). New York: Paragon.

Coleman, J. S. (1994). *Foundations of social theory*: Harvard university press.

Cooney, T. J., & Wilson, M. R. (1993). Teachers' thinking about functions: Historical and research perspective. In T. A. Romberg, E. Fennema, & T. P. Carpenter (Eds.), *Integrating research on the graphical representation of functions* (pp. 131–158). Hillsdale, NJ: Lawrence Erlbaum Associates.

Coyne, M. D., Kameenui, E. J., & Carnine, D. (2010). *Effective teaching strategies: that accommodate diverse learners Effective Teaching Strategies that Accommodate Diverse Learners* (4th ed.): Pearson Merrill Prentice Hall.

Creswell, J. W. (2012). *Qualitative inquiry and research design: Choosing among five approaches*: Sage.

Croll, P. (1986). *Systematic classroom observation*. London: Falmer Press.

Cross, D. I. (2009). Alignment, cohesion, and change: Examining mathematics teachers' belief structures and their influence on instructional practices. *Journal of Mathematics Teacher Education, 12*(5), 325–346.

Cross, D. I. (2015). Dispelling the notion of inconsistencies in teachers' mathematics beliefs and practices: A 3-year case study. *Journal of Mathematics Teacher Education, 18*(2), 173–201.

Deal, T. E. (1984). Educational change: Revival tent, tinkertoys, jungle, or carnival? *Teachers College Record, 86*(1), 124–137.

Denzin, N. (1989). *The research act: A theoretical introduction to sociological methods* (3rd ed.). Englewood Cliffs, NJ:: Prentice Hall.

Denzin, N., & Lincoln, Y. (1994). *Handbook of qualitative research*. Thousand Oaks, CA: Sage Publications.

Descartes, R. (2008). *Geometry*. (X. Yuan, Trans.). Beijing: Beijing Daxue Chubanshe [Peking University Press].

Dewey, J. (1933). *How we think: A restatement of the relation of reflective thinking to the educational process*. Lexington, MA: Health.

Dörnyei, Z. (2007). *Research methods in applied linguistics: Quantitative, qualitative, and mixed methodologies*: Oxford University Press.

Dreyfus, H. L., & Dreyfus, S. E. (1987). *Mind over machine: The power of human intuition*. New York: Free Press.

Editorial department (2008). Ruhe shiyonghao jiaofu ziliao [How to use teaching materials well]. *Zhongxuesheng Shu Li Hua (zhongxue ban) [Maths Physics & Chemistry for Middle School Students (Middle School Edition)]*, 1.

Eisner, E. W. (1991). *The enlightened eye: Qualitative inquiry and the enhancement of educational practice*. New York: MacMillan.

Ernest, P. (1989). The knowledge, beliefs and attitudes of the mathematics teacher: A model. *Journal of education for teaching, 15*(1), 13–33.

Ernest, P. (2002). *The philosophy of mathematics education*: Routledge.

Fan, L., Miao, Z., & Mok, I. A. C. (2014). How Chinese teachers teach mathematics and pursue professional development: Perspectives from contemporary international research. In

L. Fan, N.-Y. Wong, J. Cai, & S. Li (Eds.), *How Chinese teach mathematics: Perspectives from insiders* (pp. 43–70). Singapore: World Scientific Publishing Company.

Fan, L., Xiong, B., Zhao, D., Niu, W. (2018). How is cultural influence manifested in the formation of mathematics textbooks? A comparative case study of resource book series between Shanghai and England. *ZDM Mathematics Education, 50,* 787–799.

Fang, Y., & Paine, L. (2000). Challenges and dilemmas in a period of reform: Preservice mathematics teacher education in Shanghai, China. *The Mathematics Educator, 5*(1/2), 32–67.

Fantilli, R. D., & McDougall, D. E. (2009). A study of novice teachers: Challenges and supports in the first years. *Teaching and Teacher Education, 25*(6), 814–825.

Feagin, J. R., Orum, A. M., & Sjoberg, G. (1991). *A case for the case study.* Chapel Hill: University of North Caroline Press.

Feiman-Nemser, S. (1983). Learning to teach. In L. Shulman & G. Sykes (Eds.), *Handbook of teaching and policy* (pp. 150–170). New York: Longman.

Fennema, E. , & Franke, M. L. . (1992). Teachers' knowledge and its impact. In D. Groups (Ed.), *Handbook of research on mathematics teaching and learning* (pp. 147–164). New York: Macmillan.

Franke, M. L., Kazemi, E., & Battey, D. (2007). Mathematics teaching and classroom practice. In F. K. Lester (Ed.), *Second handbook of research on mathematics teaching and learning* (Vol. 1, pp. 225–256). Charlotte, NC: Information Age Publishing.

Fullan, M., & Hargreaves, A. (1992). Teacher development and educational change. In M. Fullan & A. Hargreaves (Eds.), *Teacher development and educational change* (pp. 1–9). London: Falmer Press.

Furinghetti, F. (1997). *On teachers' conceptions: From a theoretical framework to school practice.* Paper presented at the the first mediterranean conference on mathematics, Cyprus.

Furinghetti, F., & Pehkonen, E. (2002). Rethinking characterizations of beliefs. In G. Leder, E. Pehkonen, & G. Törner (Eds.), *Beliefs: A hidden variable in mathematics education?* (pp. 39–57). Dordrecht: Kluwer.

Galloway, D., Parkhurst, F., Boswell, K., Boswell, C., & Green, K. (1982). Sources of stress for classroom teachers. *National Education*(64), 166–169.

Goddard, R. D., & Goddard, Y. L. (2001). A multilevel analysis of the relationship between teacher and collective efficacy in urban schools. *Teaching and Teacher Education, 17*(7), 807–818.

Goddard, R. D., & Salloum, S. J. (2011). Collective efficacy beliefs, organizational excellence, and leadership. In K. S. Camerson & G. M. Spreitzer (Eds.), *Oxford handbook of positive organizational scholarship* (pp. 642–650). New York: Oxford University Press.

Goetz, J., & LeCompte, M. D. (1984). *Ethnography and qualitative design in educational research.* Orlando, FLA: Academic Press.

Goldsmith, L. T., Doerr, H. M., & Lewis, C. C. (2014). Mathematics teachers' learning: a conceptual framework and synthesis of research. *Journal of Mathematics Teacher Education, 17*(1), 5–36.

Goulding, M., Rowland, T., & Barber, P. (2002). Does it matter? Primary teacher trainees' subject knowledge in mathematics. *British Educational Research Journal, 28*(5), 689–704.

Graesser, A. C., & Person, N. K. (1994). Question asking during tutoring. *American Educational Research Journal, 31*(1), 104–137.

Greeno, J. G. (1997). On claims that answer the wrong questions. *Educational researcher, 26*(1), 5–17.

Gu, L. (1991). *Xuehui jiaoxue [Learning to teach].* Beijing: People's Education Press.

Gu, L., Wang, J., Yang, Y. (2014). *Xiaoben yanjiu: jiaoshi zhuanye fazhan de youli tujing [School-based study: the effective way for teacher professional development],* [PowerPoint slides]. Retrieved on February 2, 2022 from https://wenku.baidu.com/view/d13f2349fe47 33687e21aa22.html.

Gu, L., & Zhou, W. (1999). Ketang jiaoxue de guancha yu yanjiu: xue hui guancha [The observation and study of classroom teaching: learning to observe]. *Shanghai Jiaoyu [Shanghai Education], 5,* 14–18.

Gu, M. (2006). Woguo jiaoshi jiaoyu gaige de fansi [Reflection on the reform of teacher education in China]. *Jiaoshi Jiaoyu Yanjiu [Teacher Education Research], 18*(6), 3–6.

Guskey, T. R. (1986). Staff development and the process of teacher change. *Educational researcher, 15*(5), 5–12.

Hargreaves, D. H. (1999). The knowledge-creating school. *British journal of educational studies, 47*(2), 122–144.

Harrison, J., Dymoke, S., & Pell, T. (2006). Mentoring beginning teachers in secondary schools: An analysis of practice. *Teaching and Teacher Education, 22*(8), 1055–1067.

Haser, Ç., & Star, J. R. (2009). Change in beliefs after first-year of teaching: The case of Turkish national curriculum context. *International Journal of Educational Development, 29*(3), 293–302.

Hashweh, M. Z. (2005). Teacher pedagogical constructions: a reconfiguration of pedagogical content knowledge. *Teachers and Teaching, 11*(3), 273–292.

Hebert, E., & Worthy, T. (2001). Does the first year of teaching have to be a bad one? A case study of success. *Teaching and Teacher Education, 17*(8), 897–911.

Hennissen, P., Crasborn, F., Brouwer, N., Korthagen, F., & Bergen, T. (2008). Mapping mentor teachers' roles in mentoring dialogues. *Educational Research Review, 3*(2), 168–186.

Hennissen, P., Crasborn, F., Brouwer, N., Korthagen, F., & Bergen, T. (2010). Uncovering contents of mentor teachers' interactive cognitions during mentoring dialogues. *Teaching and Teacher Education, 26*(2), 207–214.

Hiebert, J., Gallimore, R., Gamier, H., Givvin, K., Hollingsworth, H., & Jacobs, J. (2003). *Teaching mathematics in seven countries: Results from the TIMSS 1999 video study. NCES 2003–013.* Washington, DC: National Center for Education Satatistics.

Hill, H. C., Sleep, L., Lewis, J. M., & Ball, D. L. (2007). Assessing teachers' mathematical knowledge: What knowledge matters and what evidence counts. In F. K. Lester (Ed.), *Second handbook of research on mathematics teaching and learning* (Vol. 1, pp. 111–155). Charlotte, NC: Information Age Publishing.

Hobson, A. J., Ashby, P., Malderez, A., & Tomlinson, P. D. (2009). Mentoring beginning teachers: What we know and what we don't. *Teaching and Teacher Education, 25*(1), 207–216.

Holliday, A. (1997). Six lessons: Cultural continuity in communicative language teaching. *Language Teaching Research, 1*(3), 212–238.

Hollingsworth, H. (1999). *Teacher professional growth: A study of primary teachers involved in mathematics professional development.* (Unpublished doctoral thesis), Deakin University, Burwood, Australia.

Hopkins, D., West, M., & Ainscow, M. (1996). *Improving the quality of education for all: Progress and challenge.* London, UK: David Fulton Publishers.

Hoy, W. K., Hannum, J., & Tschannen-Moran, M. (1998). Organizational climate and student achievement: A parsimonious and longitudinal view. *Journal of School Leadership, 8,* 336–359.

Hoy, W. K., & Sweetland, S. R. (2001). Designing better schools: The meaning and measure of enabling school structures. *Educational Administration Quarterly,* 37(3), 296–321.

Hoy, W. K., Tarter, C. J., & Hoy, A. W. (2006). Academic optimism of schools: A force for student achievement. *American Educational Research Journal, 43*(3), 425–446.

Hu, D. (2001). Jiaoshi xingxiang cong 'gongpu' dao 'zhuanjia': chuangxin jiaoxue huhuan jiaoshi zhuanyehua [The image of teacher from 'public servant' to 'expert': creative teaching calls for teacher expertise]. *Jiaoyu Fazhan Yanjiu [Exploring Education Development], 2001*(11), 50–53.

Huang, R., & Leung, F. K. (2005). Deconstructing teacher-centeredness and student-centeredness dichotomy: A case study of a Shanghai mathematics lesson. *The Mathematics Educator, 15*(2).

Huang, Z. (2016). Guanyu zhongxue shuxue "shuoti" de pingjia [About "discussing mathematical problems" in secondary schools]. *Zhongxue Shuxue Yuekan [The Monthly Journal of High School Mathematics], 2016*(1), 35–37.

Huberman, A. M., & Miles, M. B. (1984). *Innovation up close: How school improvement works.* Nueva York: Plenum.

Huberman, A. M. (1993). *The lives of teachers.* New York: Teachers College Press.

Ingersoll, R., & Kralik, J. M. (2004). *The impact of mentoring on teacher retention: What the research says. ECS research review.* Denver, CO: Education Commission of the States. Retrieved on February 2, 2022 from https://www.gse.upenn.edu/pdf/rmi/ECS-RMI-2004.pdf.

Jankvist, U. T. (2009). A categorization of the "whys" and "hows" of using history in mathematics education. *Educational Studies in Mathematics, 71*(3), 235–261.

Johnson, B., & Christensen, L. (2000). *Educational research: Quantitative and qualitative approaches.* Boston: Allyn and Bacon.

Kang, N.-H. (2008). Learning to teach science: Personal epistemologies, teaching goals, and practices of teaching. *Teaching and Teacher Education, 24*(2), 478–498.

Kelchtermans, G., & Ballet, K. (2002). The micropolitics of teacher induction. A narrative-biographical study on teacher socialisation. *Teaching and Teacher Education, 18*(1), 105–120.

Kemmis, S., Heikkinen, H. L. T., Fransson, G., Aspfors, J., & Edwards-Groves, C. (2014). Mentoring of new teachers as a contested practice: Supervision, support and collaborative self-development. *Teaching and Teacher Education, 43,* 154–164.

Kersting, N. (2008). Using video clips of mathematics classroom instruction as item prompts to measure teachers' knowledge of teaching mathematics. *Educational and Psychological Measurement, 68*(5), 845–861.

Kong, F. (2006). Shuxue jiaoshi zhuanye nengli fazhan de kunhuo ji ruogan duice [Problems and measures on the professional development of mathematics teachers]. *Hunan Jiaoyu [Hunan Education], 2006*(5), 7–11.

Kuckartz, U. (2014). *Qualitative text analysis: A guide to methods, practice and using software.* New York: Sage.

Kuhs, T. M., & Ball, D. L. (1986). *Approaches to teaching mathematics: Mapping the domains of knowledge, skills, and dispositions.* East Lansing: Michigan State University, Center on Teacher Education.

Leithwood, K., Leonard, L., & Sharratt, L. (1998). Conditions fostering organizational learning in schools. *Educational Administration Quarterly, 34*(2), 243–276.

Leung, F. K. S. (2001). In search of an East Asian identity in mathematics education. *Educational Studies in Mathematics, 47*(1), 35–51.

Leung, F. K. S. (1992). *A comparison of the intended mathematics curriculum in China, Hong Kong and England and the implementation in Beijing, Hong Kong and London.* (Unpublished doctoral dissertation), The University of London.

Li, J. (1991). *Zhongxue shuxue jiaoshi jiaoxue jiben gong jiangzuo [A lecture on the basic skills of secondary mathematics teachers].* Beijing: Beijing Shifan Daxue Chubanshe [Beijing normal university publishing group].

Li, J., & Lian, S. (1997). *Jichu jiaoyu xiandaihua jiben gong[Basic teaching skills for compulsory education (in secondary mathematics)].* Beijing: Shoudu Shifan Daxue Chubanshe [Capital Normal University Press].

Li, J., & Zhao, W. (2011). "Jiti beke": neihan, wenti yu biange celue ["Collective lesson planning": Its meaning, problems and strategies for change]. *Xibei shida xuebao (shehui kexue ban) [Journal of Northwest Normal University (Social Sciences)], 48*(6), 73–79.

Li, Q., & Ni, Y. (2009). Dialogue in the elementary school mathematics classroom: A comparative study between expert and novice teachers. *Frontiers of Education in China, 4*(4), 526–540.

Li, W. (2011). Tan "shituzhi" jiaoshi peiyang fangfa yu jiaoshi de chengzhang [Teachers' training method of "mentoring" and teachers' growth]. *Jiazhi Gongcheng [Value Engineering], 30*(21), 274–275.

Li, Y. (2002). Knowing, understanding and exploring the content and formation of curriculum materials: A Chinese approach to empower prospective elementary school teachers pedagogically. *International Journal of Educational Research, 37*(2), 179–193.

Li, Y., Zhao, D., Huang, R., & Ma, Y. (2008). Mathematical preparation of elementary teachers in China: Changes and issues. *Journal of Mathematics Teacher Education, 11*(5), 417–430.

Lian, R. (2008). Jiaoshi jiaoxue zhuanchang fazhan de xinli licheng [Mental experience of teachers' teaching expertise development]. *Jiaoyu Yanjiu [Educational research], 2008*(2), 15–20.

Lincoln, Y., & Guba, E. (1985). *Naturalistic inquiry.* Beverly Hills, CA: Sage.

Lincoln, Y. S., & Denzin, N. K. (1994). The fifth moment. In N. K. Denzin & Y. S. Lincoln (Eds.), *Handbook of qualitative research* (pp. 575–586). Thousand Oaks, CA: Sage Publications.

Lopez-Real, F., Mok, I., Leung, F., & Marton, F. (2004). Identifying a pattern of teaching: An analysis of a Shanghai teacher's lessons. In L. Fan, N.-Y. Wong, J. Cai, & S. Li (Eds.),

How Chinese learn mathematics: perspectives from insiders (pp. 382–412). Singapore: World Scientific.

Lu, X., Kaiser, G. & Leung, F.K.S. (2020). Mentoring early career mathematics teachers from the mentees' perspective—a case study from China. *International Journal of Science and Mathematics Education, 18*, 1355–1374.

Lu, X., Leung, F.K.S., & Li, N. (2021). Teacher agency for integrating history into teaching mathematics in a performance-driven context: A case study of a beginning teacher in China. *Educational Studies in Mathematics, 106*(1), 25–44.

Lumpe, A., Czerniak, C., Haney, J., & Beltyukova, S. (2012). Beliefs about teaching science: The relationship between elementary teachers' participation in professional development and student achievement. *International Journal of Science Education, 34*(2), 153–166.

Ma, Y. p., Lam, C. c., & Wong, N. y. (2006). Chinese primary school mathematics teachers working in a centralised curriculum system: a case study of two primary schools in North-East China. *Compare: A Journal of Comparative Education, 36*(2), 197–212.

MacBeath, J. (1999). *Schools must speak for themselves: The case for school self-evaluation.* London, UK: Routledge.

Macbeath, J., & Mortimore, P. (2001). *Improving school effectiveness.* UK: McGraw-Hill Education.

MacGilchrist, B., Reed, J., & Myers, K. (2004). *The intelligent school.* London, UK: Sage.

Mao, Q., & Yue, K. (2011). "Shituzhi" jiaoshi xuexi: Kunjin yu chulu [Dilemma and solution of teacher learning based on "apprenticeship"]. *Jiaoyu Fazhang Yanjiu [Research in Educational Development], 2011*(22), 58–62.

Marks, H., Louis, K. S., & Printy, S. (2000). The capacity for organizational learning: Implications for pedagogy and student achievement. In K. Leithwood (Ed.), *Understanding schools as intelligent systems* (pp. 239–265). Stamford, CT: JAI.

Merriam, S. (1988). *Case study research in education.* San Francisco: Jossey-Bass.

Merriam, S. (2009). *Qualitative Research: A Guide to Design and Implementation. Revised and Expanded from Qualitative Research and Case Study Applications in Education.* San Francisco, CA: Jossey-Bass.

Miles, M., & Hubennan, A. (1994). *Qualitative data analysis: An expanded sourcebook.* Thousand Oaks, CA: Sage.

Ministry of Education of China (2003). *Putong Gaozhong Shuxue Kecheng Biaozhun (Shiyan) [Mathematics Curriculum Standards for High Schools (Experimental)].* Beijing: People's Educational Press.

Mo, C., & He, X. (2007). Jiaoan bianxie jiqi pinggu zhibiao tixi de yanzhi [The study on the development and assessment of lesson plans]. *Jiaoyu yu Zhiye [Education and Vocation]* (23), 110–112.

Mok, I. A. C. (2006). Shedding light on the East Asian learner paradox: Reconstructing student-centredness in a Shanghai classroom. *Asia Pacific Journal of Education, 26*(2), 131–142.

Mouza, C. (2009). Does research-based professional development make a difference? A longitudinal investigation of teacher learning in technology integration. *Teachers College Record, 111*(5), 1195–1241.

Moyer, P. S., & Milewicz, E. (2002). Learning to question: Categories of questioning used by preservice teachers during diagnostic mathematics interviews. *Journal of Mathematics Teacher Education, 5*(4), 293–315.

Mullis, I. V., Martin, M. O., Foy, P., & Arora, A. (2012). *TIMSS 2011 international results in mathematics.* Chestnut Hill, MA: TIMSS & PIRLS International Study Center, Boston College.

Nickols, F. (2000). What is in the world of work and working: Some implications of the shift to knowledge in work. In J. W. Cortada & J. A. Woods (Eds.), The knowledge management yearbook, 2000–2001 (pp. 3–11). Boston, MA: Butterworth Heinemann.

Nonaka, I., & Takeuchi, H. (1995). *The knowledge-creating company: How Japanese companies create the dynamics of innovation.* New York, NY: Oxford University Press.

Novak, D., & Knowles, J. G. (1992). *Life histories and the transition to teaching as a second career.* Paper presented at the the the annual meeting of the American Educational Research Association,, Chicago, IL.

Opfer, V. D., & Pedder, D. (2011). Conceptualizing teacher professional learning. *Review of educational research, 81*(3), 376–407.

Organisation for Economic Co-operation and Development (2012). *PISA 2012 Results in Focus: What 15-year-olds know and what they can do with what they know.* Geneva: OECD Programme for International Student Assessment.

Organisation for Economic Co-operation and Development (2015). *Results from TALIS 2013–2014: Shanghai (China) country note.* Retrieved on February 9, 2019 from www.oecd.org/education/school/TALIS-2014-country-note-Shanghai.pdf.

Organisation for Economic Co-operation and Development. (2005). *Teachers matter: Attracting, developing and retaining effective teachers,* Pairs: OECD Publishing.

Orland-Barak, L. (2014). Mediation in mentoring: A synthesis of studies in teaching and teacher education. *Teaching and Teacher Education, 44,* 180–188.

Pajares, M. F. (1992). Teachers' beliefs and educational research: Cleaning up a messy construct. *Review of educational research, 62*(3), 307–332.

Patton, M. Q. (2002). *Qualitative research and evaluation methods.* Thousand Oaks, Calif: Sage.

Pedder, D. (2006). Organizational conditions that foster successful classroom promotion of Learning How to Learn. *Research papers in education, 21*(02), 171–200.

Petrou, M., & Goulding, M. (2011). Conceptualising teachers' mathematical knowledge in teaching. In T. Rowland & K. Ruthven (Eds.), *Mathematical knowledge in teaching* (pp. 9–25): Springer.

Philipp, R. A. (2007). Mathematics teachers' beliefs and affect. In F. K. Lester (Ed.), *Second handbook of research on mathematics teaching and learning: A project of the National Council of Teachers of Mathematics* (Vol. 1, pp. 257–315). Charlotte, NC: Information Age Publishing.

Postholm, M. B. (2012). Teachers' professional development: a theoretical review. *Educational research, 54*(4), 405–429.

Raths, J. (2001). Teachers' beliefs and teaching beliefs. *Early Childhood Research & Practice, 3*(1).

Raths, L., Harmin, M., & Simon, S. (1987). Selections from 'values and teaching'. In J. P. F. Carbone (Ed.), *Value theory and education* (pp. 198–214). Malabar, FL: Robert E. Krieger.

Raymond, A. M. (1997). Inconsistency between a beginning elementary school teacher's mathematics beliefs and teaching practice. *Journal for research in mathematics education,* 550–576.

Reynolds, A. (1992). What is competent beginning teaching? A review of the literature. *Review of educational research, 62*(1), 1–35.

Richardson, V. (1996). The role of attitudes and beliefs in learning to teach. In J. Sikula (Ed.), *Handbook of research on teacher education* (2 ed., pp. 102–119). New York: Macmillan.

Richardson, V. (2003). Preservice teachers' beliefs. In J. Raths & A. C. McAninch (Eds.), *Teacher beliefs and classroom performance: The impact of teacher education* (Vol. 6, pp. 1–22). Greenwich: Information Age.

Rokeach, M. (1973). *The nature of human values.* New York: Free press.

Rosenholtz, S. J., Bassler, O., & Hoover-Dempsey, K. (1986). Organizational conditions of teacher learning. *Teaching and Teacher Education, 2*(2), 91–104.

Rowland, T. (2013). The knowledge quartet: the genesis and application of a framework for analysing mathematics teaching and deepening teachers' mathematics knowledge. *Sisyphus-Journal of Education, 1*(3), 15–43.

Rowland, T., Turner, F., & Thwaites, A. (2014). Research into teacher knowledge: a stimulus for development in mathematics teacher education practice. *ZDM—Mathematics Education, 46*(2), 317–328.

Salleh, H., & Tan, C. H. P. (2013). Novice teachers learning from others: Mentoring in Shanghai schools. *Australian Journal of Teacher Education, 38*(3), 152–165.

Sampson, R. J., Morenoff, J. D., & Earls, F. (1999). Beyond social capital: Spatial dynamics of collective efficacy for children. *American sociological review*, 633–660.

Schein, E. H. (2006). *Organizational culture and leadership* (Vol. 356). Hoboken, NJ: Wiley.

Schoenfeld, A. H. (2002). How can we examine the connections between teachers' world views and their educational practices. *Issues in Education, 8*(2), 217–227.

Schoenfeld, A. H. (2010). *How we think: A theory of goal-oriented decision making and its educational applications.* New York: Routledge.

Seashore Louis, K., Kruse, S., & Marks, H. M. (1996). School-wide professional community: Teachers' work, intellectual quality, and commitment. In F. W. Newman & Associates (Eds.), *Authentic achievement: Restructuring schools for intellectual quality* (pp. 179–203). San Francisco, CA: Jossey-Bass.

Senge, P. M. (1990). *The fifth discipline: The art and practice of the learning organization.* New York: Doubleday.

Sfard, A. (1998). On two metaphors for learning and the dangers of choosing just one. *Educational researcher, 27*(2), 4–13.

Shi, L. (2002). Woguo jiaoshi xiaoben peixun yanjiu zongshu [Literature review on school-based training for teachers in China]. *Chengren Gaodeng Jiaoyu [Journal of Adult Higher Education], 2002*(5), 22–26.

Shulman, L. S., & Grossman, P. (1988). *Knowledge growth in teaching: A final report to the Spencer Foundation.* Stanford, CA: Stanford University.

Shulman, L. S. (1986). Those who understand: Knowledge growth in teaching. *Educational researcher, 15*(2), 4–14.

Shulman, L. S. (1987). Knowledge and teaching: Foundations of the new reform. *Harvard educational review, 57*(1), 1–23.

Simons, H. (2009). *Case study research in practice.* London: SAGE.

Skott, J. (2009). Contextualising the notion of 'belief enactment'. *Journal of Mathematics Teacher Education, 12*(1), 27–46.

Song, H. (2012). Xinjiaoshi zhuanye fazhan: Cong shitudaijiao zouxiang zhuanye xuexi qunti [New teachers' professional development: From mentoring to professional learning community].*Waiguo Jiaoyu Yanjiu [Studies in Foreign Education], 39*(262), 77–84.

Sowder, J. (1998). *Middle grade teachers' mathematical knowledge and its relationship to instruction: A research monograph.* New York: State University of New York Press.

Sowder, J. (2007). The mathematical education and development of teachers. In F. K. Lester (Ed.), *Second handbook of research on mathematics teaching and learning* (Vol. 1, pp. 157–233). Charlotte, NC: Information Age Publishing.

Speer, N. M. (2005). Issues of methods and theory in the study of mathematics teachers' professed and attributed beliefs. *Educational Studies in Mathematics, 58*(3), 361–391.

Stake, R. (1988). Case study methods in educational research: Seeking sweet water. In R. M. Jaeger (Ed.), *Complementary methods for research in education.* Washington, DC: American Educational Research Association.

Stake, R. (1994). Case Studies. In N. K. Denzin & Y. S. Lincoln (Eds.), *Handbook of qualitative research* (pp. 575–586). Thousand Oaks, CA: Sage Publications.

Stigler, J. W., & Hiebert, J. (1999). *The teaching gap: Best ideas from the world's teachers for improving education in the classroom.* New York:: The Free Press.

Stipek, D. J., Givvin, K. B., Salmon, J. M., & MacGyvers, V. L. (2001). Teachers' beliefs and practices related to mathematics instruction. *Teaching and Teacher Education, 17*(2), 213–226.

Tatto, M. T., Schwille, J., Senk, S., Ingvarson, L., Peck, R., & Rowley, G. (2009). *Teacher Education and Development Study in Mathematics (TEDS-M): conceptual framework: policy, practice, and readiness to teach primary and secondary mathematics*: International Association for the Evaluation of Educational Achievement (IEA).

Thompson, A. G. (1991). *The development of teachers' conceptions of mathematics teaching.* Paper presented at the the thirteenth annual meeting of the North American Chapter of the International Group for the Psychology of Mathematics Education (Vol. 2), Blacksburg, VA.

Thompson, A. G. (1992). Teachers' beliefs and conceptions: A synthesis of the research. In D. Grouws (Ed.), *Handbook of research on mathematics teaching and learning* (pp. 127–146). New York: Macmillan.

Timperley, H., & Alton-Lee, A. (2008). Reframing teacher professional learning: An alternative policy approach to strengthening valued outcomes for diverse learners. *Review of research in education, 32*(1), 328–369.

Tirosh, D., & Even, R. (2007). *Teachers' knowledge of students' mathematical learning: An examination of commonly held assumptions.* Paper presented at Mathematics knowledge in teaching seminar series: Conceptualising and theorizing mathematical knowledge for teaching (Seminar1). Cambridge, MA: University of Cambridge.

Tschannen-Moran, M., Bankole, R. A., Mitchell, R. M., & Moore Jr, D. M. (2013). Student Academic Optimism: a confirmatory factor analysis. *Journal of Educational Administration, 51*(2), 150–175.

Tschannen-Moran, M., Parish, J., & DiPaola, M. (2006). School climate: The interplay between interpersonal relationships and student achievement. *Journal of School Leadership, 16*(4), 386.

Tschannen-Moran, M., Salloum, S. J., & Goddard, R. (2014). Context matters: The influence of collective beliefs and shared norms. In H. Fives & M. G. Gill (Eds.), *International* handbook of research on teacher beliefs (Educational psychology handbook) (pp. 301–313). New York: Routledge.

Tschannen-Moran, M., & McMaster, P. (2009). Sources of self-efficacy: Four professional development formats and their relationship to self-efficacy and implementation of a new teaching strategy. The Elementary School Journal, 110(2), 228–245.

Tu, R., & Song, X. (2006). Zhongguo shuxue jiaoxue de ruogan tedian [Several characteristics of mathematics teaching in second schools]. Kecheng Jiaocai Jiaofa [Curriculum, Teaching Material and Method], 26(2), 43–46.

Turner, F., & Rowland, T. (2011). The Knowledge Quartet as an organising framework for developing and deepening teachers' mathematics knowledge. In T. Rowland & K. Ruthven (Eds.), Mathematical knowledge in teaching (pp. 195–212). New York: Springer.

Turner, J. C., Warzon, K. B., & Christensen, A. (2011). Motivating mathematics learning changes in teachers' practices and beliefs during a nine-month collaboration. American Educational Research Journal, 48(3), 718–762.

Tzanakis, C., & Arcavi, A. (2000). Integrating history of mathematics in the classroom: An analytic survey. In J. Fauvel & J. v. Maanen (Eds.), History in mathematics education (pp. 201–240). The ICMI Study. Dordrecht: Kluwer Academic Publishers.

Van Lier, L. (1988). The classroom and the language learner. London: Longman.

Van Maanen, J. (1979). The fact of fiction in organizational ethnography. Administrative Science Quarterly, 24(4), 539–550.

Van Zoest, L. R., Jones, G. A., & Thornton, C. A. (1994). Beliefs about mathematics teaching held by pre-service teachers involved in a first grade mentorship program. Mathematics Education Research Journal, 6(1), 37–55.

Wan, L. (1999). Chuzhong jiaoshi kaizhan "sanzi yihua yiji" jibengong xunlian hen youbiyao [The need of basic training for junior middle school teachers]. Nanchang Jiaoyu Xueyuan Xuebao [Journal of Nanchang College of Education] (4), 15–18.

Wang, H. (2013). Shuxue haoke de biaozheng: beijing zongshu he liangbiao sheji [Characterisation of a good mathematical lesson: the review of background and the design of evaluation]. Shuxue Jiaoxue [Mathematical Teaching], 8, 1–4.

Wang, J. (2009). Cong "shitudaijiao" dao "tuanduichengzhang"—Jiyu Shanghaishi bufen xinjiaoshi zhuanye chengzhang diaoyan de yanjiu [Investigation and exploration on new teachers' professional growth]. Jiaoyu Fazhan Yanjiu [Research in Educational Development], 2009(24), 67–71.

Wang, J. (2013). Jiaoshi jiben gong de neihan ji fazhan tujing [the meaning and development of teacher basic skills]. Liaoning jiaoyu[Liaoning Education], 2013(5), 5–7.

Wang, S. (2006). Jiti beike budengyu tongyi jiaoan [Collective lesson planning is not equal to unifying lesson plans]. Jiangsu Jiaoyu (jiaoyu guanli ban) [Educational Management of Jiangsu Education] (3), 25–25.

Wang, T., & Cai, J. (2007). Chinese (Mainland) teachers' views of effective mathematics teaching and learning. ZDM—Mathematics Education, 39(4), 287–300.

Wang, X. (2013). HPM yu chuzhong shuxue jiaoshi de zhuanye fazhang [Teacher's professional development promoted by HPM: the case of a junior high school mathematics teacher in Shanghai]. Shuxue Jiaoyu Xubao [Journal of mathematics education], 22(1), 18–22.

Wei, Z., & Bao, C. (2009). Shanghaishi zhongxiaoxue jiaoshi xueli tisheng jiaoyu zhuangkuang diaocha yu fenxi [The investigation and analysis on the primary and middle school Teachers' 'Educational background upgrade education']. *Jiaoyu Fazhan Yanjiu [Exploring Education Development], 2009*(15–16), 66–69.

Weston, T. L., Kleve, B., & Rowland, T. (2012). Developing an online coding manual for the Knowledge Quartet: An international project. *Proceedings of the British Society for Research into Learning Mathematics, 32*(3).

Wheatley, K. F. (2002). The potential benefits of teacher efficacy doubts for educational reform. *Teaching and Teacher Education, 18*(1), 5–22.

Wong, N.-Y. (2002). Conceptions of doing and learning mathematics among Chinese. *Journal of Intercultural Studies, 23*(2), 211–229.

Wong, N.-Y., Ding, R., & Zhang, Q. P. (2016). From classroom environment to conception of mathematics. In R. B. King & A. B. I. Bernardo (Eds.), *The Psychology of Asian Learners* (pp. 541–557). Singapore: Springer.

Wong, N.-Y., Han, J., & Lee, P. Y. (2004). The mathematics curriculum: toward globalization or westernization? In L. Fan, N. Y. Wong, J. Cai, & S. Li (Eds.), *How Chinese learn mathematics: Perspectives from insiders* (pp. 27–70). Singapore: World Scientific.

Woods, P., Jeffrey, B., Troman, G., & Boyle, M. (1997). *Restructuring schools, reconstructing teachers: Responding to change in the primary school.* Buckingham: Open University Press.

Woolfolk Hoy, A., & Burke-Spero, R. (2005). Changes in teacher efficacy during the early years of teaching: A comparison of four measures. Teaching and Teacher Education, 21, 343–356.

Woolfolk Hoy, A., Hoy, W. K., & Davis, H. A. (2009). Teachers' self-efficacy beliefs. In K. Wentzel & A. Wigfield (Eds.), *Handbook of motivation in school* (pp. 627–655). Mahwah, NJ: Lawrence Erlbaum.

Wu, Y. (2012). *The examination system in China: The case of Zhongkao mathematics.* Paper presented at the Selected Regular Lectures from the 12th International Congress on Mathematical Education.

Xu, H. (2010). Jiaoyanzu he beikezu de gongneng yu yunxing [The function and operation of teaching research group and lesson preparation group]. *Renmin Jiaoyu [People's Education]* (22), 51–53.

Xu, Y. (2004). Cong jichu jiaoyu kecheng gaige xuyao chufa chongxin sikao jiaoshi jiaoxue jibengong [Rethinking teachers' basic teaching skills from the necessities of the basic educational curriculum reform]. *Kecheng Jiaocai Jiaofa Curriculum, [Teaching Material and Method], 24*(2), 73–79.

Yan, D., & Duan, W. (2009). Cong "shi tu zhi" dao "xue tu zhi" qun zhi [From one-to-one mentoring to group mentoring]. *Bianji xuekan [Editors Monthly], 2009*(4), 80–82.

Yang, J. (2015). *Lun shuxue jiaoshi de jibengong [Mathematics teaching basic skills].* (Unpublished master's dissertation), Central China Normal University.

Yang, Q. (1995). *Kunhuo yu jueze: 20 shiji de xin jiao xue lun [Confusion and choices: new teaching theories in the 20th century].* Jinan, Shandong: Shangdong jiaoyu chu ban she [Shandong education press].

Yin, R. (1994). *Case study research: Design and methods* (2 ed.). Thousand Oaks, CA: Stage.

Yin, R. (2010). *Qualitative research from start to finish*: Guilford Press.

Zack, M. H. (2000). Managing organizational ignorance. In J. W. Cortada & J. A. Woods (Eds.), *The knowledge management yearbook, 2000–2001* (pp. 353–373). Boston, MA: Butterworth Heinemann.

Zhang, D. (2010). Jianshe zhongguo tese de shuxue jiaoyu lilun [Constructing mathematics education theory with Chinese characteristics]. *Shuxue jiaoxue [Mathematical Teaching], 2010*(1), 0–7.

Zhang, M. (2012). Qiantan due "xiaoben zuoye" de jidian sikao [Reflections on the "school-based exercises"]. *Zhongxue Jiaoxue Canker [Reference for Middle School Teaching]* (3), 125–125.

Zhang, Q., & Wong, N.-Y. (2014). Beliefs, knowledge and teaching: A series of studies about Chinese mathematics teachers. In L. Fan, N.-Y. Wong, J. Cai, & S. Li (Eds.), *How Chinese teach mathematics: Perspectives from insiders* (pp. 457–492). Singapore: World Scientific Publishing Company.

Printed in the United States
by Baker & Taylor Publisher Services